Speech Technology at Work

IFS

SPEECH TECHNOLOGY AT WORK

Jack Hollingum
Graham Cassford

Springer-Verlag Berlin Heidelberg GmbH

Jack Hollingum
IFS (Publications) Ltd
35–39 High Street
Kempston
Bedford MK42 7BT
UK

Graham Cassford
Logica Energy & Industry Systems Ltc
64 Newman Street
London
W1A 4SE
UK

British Library Cataloguing in Publication Data

Hollingum, Jack
    Speech technology at work.
    1. Manufacturing industries. Applications of speech recognition by computer
    systems
    I. Title II. Cassford, Graham, 1957–

ISBN 978-3-662-13014-8   ISBN 978-3-662-13012-4 (eBook)
DOI 10.1007/978-3-662-13012-4

© **1988** Springer-Verlag Berlin Heidelberg
Ursprünglich erschienen bei Springer-Verlag Berlin Heidelberg New York Tokyo **1988**

Phototypeset by Fleetlines Typesetters, Southend-on-Sea

Contents

Graham Cassford gained a BSc Hons from Portsmouth Polytechnic in 1980 and a PhD from Reading University in 1983. He began his career with British Robotic Systems Ltd as a software engineer concerned with the development of Industrial Vision Processing Systems. In 1985 he joined Logica Energy and Industry Systems Ltd as a senior consultant within the Industrial Systems Group of the company. He is responsible for the design and implementation of shop-floor computer systems, specialising in the application of speech- and vision-based technologies.

Jack Hollingum is a distinguised technical author and journalist, and a former editor of *Sensor Review* magazine. His previous publications include *Machine Vision: The Eyes of Automation* (1984); *The Machine Vision Sourcebook* (1986); *Implementing an Information Strategy in Manufacture* (1987); and *Implementing Total Quality* (1987), all published by IFS.

Preface

Speech technology – the use of speech as a means of sending information to, and receiving information from computer systems – has been in use as a research tool for many years. Only recently has it begun to move out of the laboratory and into commercially worthwhile applications, first with compressed and synthesised spoken messages, then with computer recognition of spoken messages, and today with diverse applications involving both recognition and reproduction of human speech.

We have written this book because we believe the technology has now advanced to the point where many more applications of voice recognition and response are both feasible and economically attractive. Computers that can understand everyday speech are still a distant prospect, but provided the limitations of present day equipment are clearly understood there is much that can be achieved with it.

Our aim is to show, in non-technical language, what is now possible with the help of speech technology. The text includes many examples of current applications in industry, commerce and other fields, and we have selected five current industrial applications combining speech recognition and response for more detailed attention. Industrial cases have been chosen both because we see industry as an important growth area for speech applications in the next few years, and because it presents some of the greatest difficulties in speech recognition – if you can make it work in industry, then you can make it work almost anywhere.

Included in the book as an appendix are brief details of companies offering equipment or services connected with speech technolgy. It is based on information supplied by those compaies and makes no claim either to completeness or to endorsing the products and services mentioned. Nevertheless we hope that it will be of some help to anybody wishing to explore further the opportunities for introducing speech technology.

A book like this inevitably draws on the knowledge and experience of a great many people, and it would be impossible to name or even to identify all of them. Several people, though, have given personal advice and help which we would like to acknowledge without in any way implicating them in any errors which may have crept into the text.

We have drawn on the experience of a number of people in Logica. In particular Dr Jeremy Peckham, who is an international authority on speech recognition, has given a great deal of advice on the subject-matter of the book, as well as reading and commenting on the final draft. Chapters 2, 5, 6 and 8 in particular make considerable use of his expertise.

Chris Wheddon, head of speech and language processing at the British Telecom Research Laboratories, gave a fascinating insight into work which should soon bear fruit in a number of important areas. Raj Gunawardana of the radio and speech systems group at Texas Instruments was helpful in reviewing developments in voice response and speaker independence.

For information relating to the industrial case studies we are deeply indebted to Jaguar Cars Ltd., Caterpillar (UK) Ltd.; to Kapul Gill of Austin Rover for his contribution on the Rover Inspection system; to Alan Patrick of Rolls-Royce for detailed information on the voice-controlled inspection system at the company's Precision Casting Facility; and to Ford Motor Company and Computer Gesellschaft Konstant for information on the Cologne Distribution Centre. Other application details have come from many sources, but particularly from the companies which have also supplied details of their products and services – our grateful thanks to them.

1 INTRODUCTION

A T JAGUAR Cars, an inspector on a vehicle quality surveillance audit stands listening to an instruction asking him to check the boot lid. He observes that its alignment relative to surrounding panels is correct and speaks into a microphone: 'OK'. At an adjacent audit stand an inspector who is working on another car is asked to look at the boot lock and finds that it does not work. She speaks: 'Fault Inoperative Alert Final Complete'. As she speaks, the information is automatically transmitted by radio to a computer which processes it and records it in a data bank for quality analysis. Shortly afterwards a message is broadcast on a television monitor in the final assembly zone where boot locks are fitted to the car, warning operators to check the functioning of locks. A message is also transmitted to the downstream rectification area with an instruction to seek out and replace the faulty lock, and also to check the locks on other cars in the present audit batch going down the line.

At the parts distribution centre of the Ford Motor Company in Cologne, operators identify and weigh packages, combine them into loads for different destinations, produce delivery notes and carry out various other tasks. They communicate with the computer which generates lists and invoices by using different methods for different purposes. Walking around the shipping area floor they identify packages to the computer by speaking its number into a headset microphone. They put the package on the weighing machine, and say 'Go' to the computer, which records the weight directly. Occasionally they go to the keyboard and enter the details for complete loads – destination, lorry number plate and so on.

At Rolls-Royce's precision casting facility in Derby inspectors are ultrasonically inspecting turbine blades several times faster than previously by speaking the results of their measurements into a computer while having their hands free to carry out the difficult manipulation of the blades and their eyes free to read the measurements from an oscilloscope. Because the information goes straight into the computer instead of having to be keyed in later, analysis of a batch of blades which used to take engineers about a day and a half can now be completed in 15 minutes.

Fig. 1.1. Caterpillar test operator using speech technology in a hostile environment.

In a very different field of application, there are expanding facilities for stockmarket investors to obtain up-to-date information by telephone, using a service which combines digitised and compressed speech with the use of a dual tone multi-frequency – DTMF – telephone (usually called 'Touch-Tone' in the USA after the trade-mark of AT&T), or with a tone-pad and any telephone. A typical UK example is the Financial Times Cityline. Having dialled the Cityline number the caller is given the option of a list of financial reports from which details can be requested, or prices of individual shares can be obtained. For a subscription fee it is also possible to keep a portfolio of up to 20 shares in the FT computer and obtain at any time their real-time prices. An extension of the system being planned will make it possible to bypass the need for a DTMF telephone or tone-pad – the caller will be able to speak the required service and a speech recognition system in the receiving computer will process the requests.

Pop record enthusiasts in the UK can hear any number of tracks from each of the top five Gallup Chart LPs and other selected tracks from singles and albums, by making a telephone call to a computer system run by the Virgin group. The caller can control what is heard by speaking commands to the computer. The choice of words is small but adequate – 'Yes', 'No', 'Repeat' and 'Change'. The caller is first prompted to speak these four words, and the system stores the sounds as 'templates' against which future spoken words will be compared. It then takes the caller through a series of options to get to the desired record tracks.

These are just a few of the ways in which human speech is being used as a way of talking with computers – both sending and receiving information.

Speech is the most natural method by which people communicate with each other, so that if computers can be made to 'understand' human speech and to speak in response, it will be a significant advance in the ease with which we can use computers in everyday life.

We are still a very long way from the time when we can carry on a conversation with a computer in the same way that we would with a colleague, but great steps forward have been taken in recent years in the computer recognition of speech, and particularly in the compression and synthesis of speech. As a result there are now rapidly growing numbers of industrial and commercial applications of speech technology, especially in the USA, and more and more companies are finding them an aid to profitability.

The difficulty, at this stage of a fast-changing technology, is in knowing what types of applications are worth pursuing today, which are the areas to keep an eye on for opportunities in the near future, and which are the pipe-dreams which may materialise in another 10 or 20 years.

In this book our aims are:

- To explain enough about speech technology, in non-technical language, for you to find your way through the jungle of jargon in sales literature and technical press articles on the subject, and make your own assessment of the claims of its advocates.
- To indicate the areas where speech technology is already being applied successfully and where it shows interesting potential for the near future.
- To help you to take the first steps towards implementing a speech technology application.

This book is written primarily for senior management in business, who will need to know about speech systems, which we see as an important growth area in the next few years. The scope for applications is very wide, and we shall draw our examples from many areas of industry and commerce.

Let's start with a brief summary of the area which is covered by practical speech technology applications today.

SPEECH RECOGNITION

This is a developing area and many things are possible under controlled laboratory conditions which are not yet rugged enough for the ordinary world of business and industrial applications. Equipment that you can go out and buy today is no more than a switching device. You speak a word or a phrase and the system tries to match the sound with a collection of sounds held in its memory. If it finds a matching sound it makes a predetermined response. It may speak a synthesised message. It may print a message on the computer screen. Or it may carry out a complicated series of instructions.

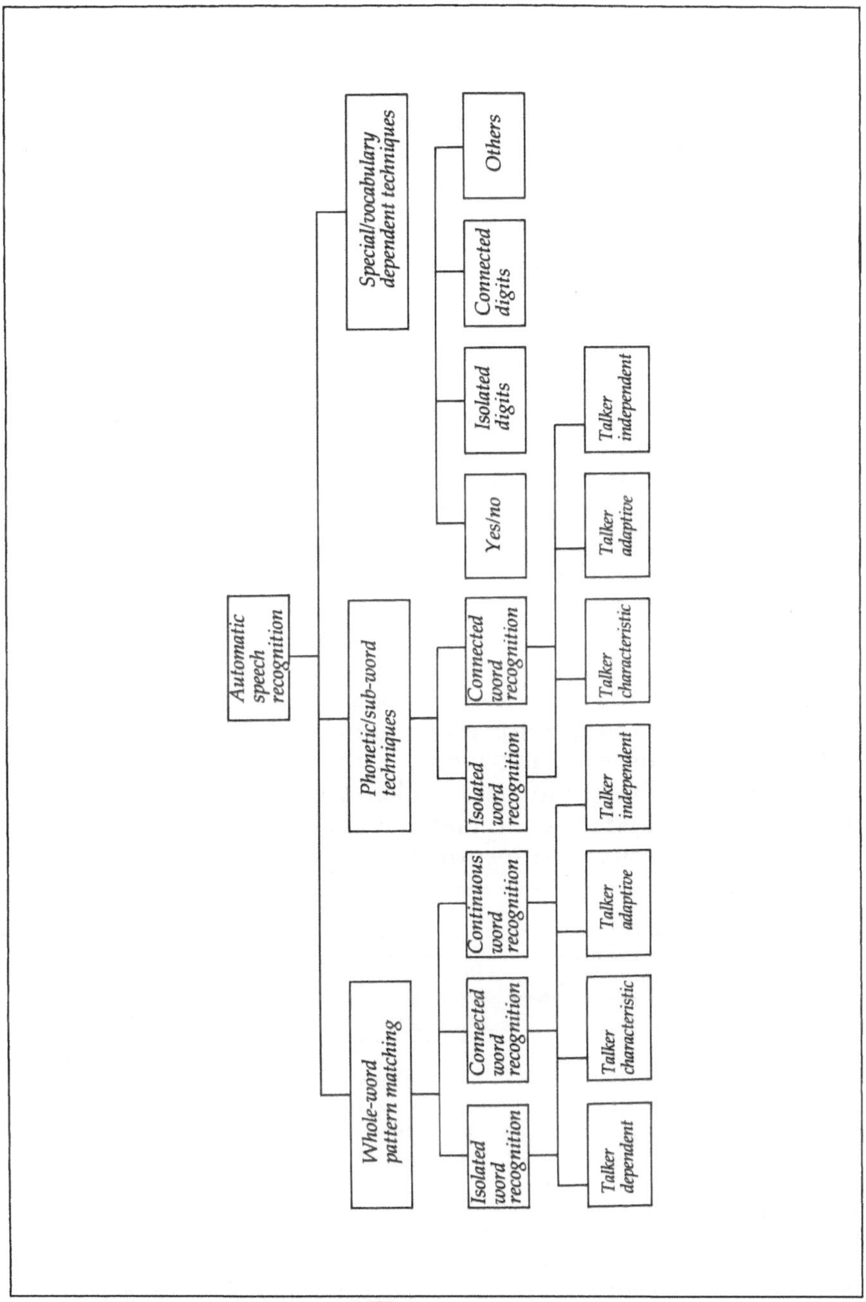

Fig. 1.2. *Different approaches to automatic speech recognition*

There is a great deal of refinement in the technology of matching sounds to stored signals and, as Fig. 1.2 shows, there are many different ways of carrying out the task, but what can at times appear to be the result of uncanny intelligence is in reality no more than a pattern-matching device, though there may be considerable subtlety in the method of pattern matching.

Three different types of recognition can be distinguished in current applications:

- *Speaker-dependent recognition.* The person who will be using the system first 'teaches' it by speaking the vocabulary of words or short phrases that will be used in the application. In use, the words or phrases will probably have to be spoken separately, with a short pause between each, or at best may be connected but separately articulated. Present-day systems will not recognise everyday spoken conversation.
- *Speaker-independent recognition.* For a very small vocabulary of perhaps 15 words, systems can respond fairly reliably to many different speakers. This is valuable, though not absolutely essential for something like a telephone enquiry service, but the speaker has to be restricted to using probably the 10 digits together with 'yes', 'no' and maybe a few other words. The usual method of working is for the system to store a large variety of sound patterns for each word, with different accents, and to search for the best fit for each spoken word.
- *Speaker identification.* Everybody's voice is slightly different, and this fact can be used in security systems to check the identity of a

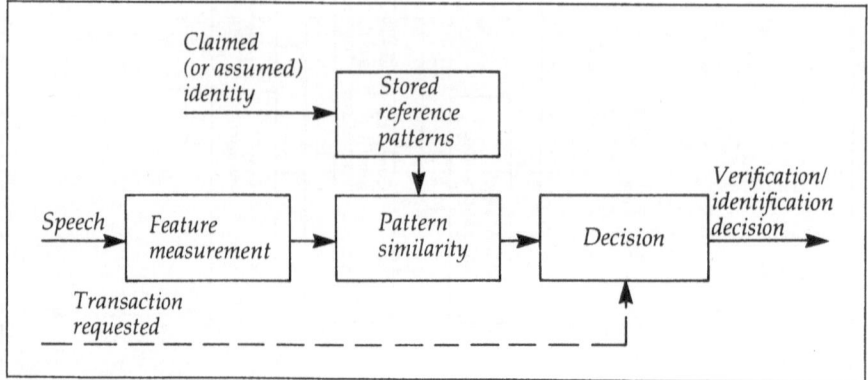

Fig. 1.3. Block diagram of a speaker identification system. Pattern matching is used to verify the identity of the speaker

speaker by asking him or her to repeat a phrase and comparing the response with a stored voice print, as shown diagrammatically in Fig. 1.3. This is not an absolutely secure method, though for most purposes it is very reliable. The threshold needs to be set high enough to prevent other speakers from gaining access by imitating the voice, which means that the authentic speaker may occasionally be rejected.

Speech output and recognition are largely distinct technologies and can be used independently of each other. In many industrial speech recognition applications, though, speech output is a natural partner in a two-way spoken communication system.

SPEECH OUTPUT

This covers several types of applications:

- *Reproduction of fixed messages.* An ordinary tape-recording of the human voice is the simplest way of playing back a message, and is the basis of the everyday telephone answering machine. However, this method is very limited in scope, so ways have been found to convert speech to digital signals which can be used by a computer, and to simplify and compress the information so that it occupies a much smaller space in the computer memory. As a result:
    - Spoken telephone messages can be stored automatically and played back when required in a much more flexible way than is possible with a telephone answering machine. This capability has given rise to a voice equivalent to electronic mail.
    - Messages can be stored permanently on a read-only chip and replayed selectively by the computer. A well-known early example was the speech chip used by Austin Morris for the Maestro, which reminded the driver to fasten the seat belt.
- *Production of variable messages.* The earliest example of this was TIM, the speaking clock, introduced in the 1930s. This used what today appears a primitive combination of separate analogue recordings for the spoken messages like 'At the third stroke it will be . . .'; and the spoken number elements for the time in hours, minutes and seconds. Today the digitised elements of the message can be assembled much more simply within the computer.

- *Generation of synthesised speech from text,* either held in the computer or read automatically from a typescript or a book by using optical character recognition. This introduces a new set of problems, to which a variety of solutions are available and will be reviewed later in this book.

WHAT MAKES SPEECH TECHNOLOGY WORTH INVESTIGATING?

People have probably been talking as long as there have been other people to talk to, and it is difficult to imagine a world without speech. A baby can make its wants known very effectively during its first few months, but learning to talk is a tremendous advance in its relationship with the outside world. It is hardly surprising, therefore, that a great amount of effort has gone into the development of man–machine interfaces which make use of speech.

The fact that speech is a natural means of communication for people is a powerful argument in favour of speech recognition and synthesis, but it is important not to overstate this argument. Synthetic speech, if of poor quality, can be annoying as well as confusing, and if of good quality can mislead the hearer into thinking he or she is dealing with another person. Similarly, the inadequacies of speech recognition, if it is badly applied, can lead to frustration and rejection of the whole method.

Realistically, the biggest scope for speech systems is where the user needs to have eyes and hands free, or where communication must be by telephone. The first of these conditions describes very many inspection situations. An inspector wants to be inspecting – looking for possible faults. Stopping to write notes, or to enter them at a keyboard, is a distraction. If the inspection also involves clambering over a vehicle, a ship's hull or a fabricated structure, a note pad may be an inconvenience and a hand-held terminal an impossibility.

Flying a military aeroplane, and particularly a helicopter, demands very close and continuous observation as well as the use of hands and feet, so there is inevitably a considerable defence interest in speech systems, both for input and output.

But there are many other, less exacting, tasks where eyes and hands are busy and some other means of communication can be of assistance. A computer-aided design terminal, for example, is a very complicated device, making use of a keyboard, a tablet and possibly a light-pen, and it can be a great help if the draughtsman can simply speak certain commands instead of having to look away from the monitor screen and search for them on the tablet display.

Any situation requiring the use of a telephone is an obvious opportunity for speech recognition as well as synthesis. In the USA, DTMF keypads are used in telephone access to computer data banks using speech synthesis, and similar devices can be added to pulse dial telephones, but automatic speech recognition offers a more flexible input, though limited at present by the capability of speaker-independent recognition or the need to go through a routine of teaching the equipment to recognise the speaker's voice.

Situations where speech communication is a virtual necessity because of demands made on the other senses are relatively few, but there are many potential applications where some gain in efficiency can be achieved by avoiding the distraction of writing or keyboard entry.

More important than this greater convenience, though, is the fact that information is captured immediately, in real time, and in a computer-usable form. If an inspector goes out and makes notes on a piece of equipment, and then returns to his or her department to report, there is a loss of hours or even days before anything is done as a result of the report. The report itself is hand-written or typed, and if any statistical analysis is to be made of the results of, say a week's output, the information must then be collected together and keyed into a computer terminal or analysed manually. In the Jaguar application quoted above, the inspector's findings are transmitted immediately to the people concerned, both upstream and down-stream on the production line, and the information is stored in a suitable form for longer-term statistical analysis.

Immediate reporting means that faults can be corrected more speedily and with greater certainty, adverse trends can be detected and rectified before they reach the stage of causing failure, and the morale and team spirit of everybody involved in the task are enhanced.

Besides the promptness of speech input, the method eliminates the danger of recording errors – ticking the wrong box or pressing the wrong key – because any important information can be automatically checked and played back to the operator for confirmation. It also avoids the inherent danger of transcription errors where information is later keyed into a computer.

A voice input/output headset can be worn in almost any environment, where it would be difficult to use any other communication medium – for example, by the driver of an airport kerosene tanker wearing oily gloves in freezing conditions, or a surgeon requiring information from a computer monitoring the patient's condition.

LIMITATIONS

Speech communication has certain limitations, some of which have been mentioned already. Though it may be more convenient than manual note-taking, it can be slower, especially if information is played back, and so can be annoying to the user. In situations where the user has easy access to a keyboard the real-time data transfer facilities are the same as for speech input and the cost is less than for a speech system.

Just as manual data input is subject to errors, so is speech input. A small proportion of voice input messages can be misrecognised – typically between 1% and 5% depending on the system, the operating conditions, the size of the vocabulary being used, and the speaker. Safeguards can be incorporated, including intercepting obvious errors and using synthesis to play back a 'Did you say . . .?' The extent to which safeguards are employed will depend on how critical the information is, and this will be a factor in any decision about using speech technology.

Very stressful physical conditions can influence the speaker's voice and affect the quality of recognition, though this is probably of more significance in military than in industrial circumstances. Environmental noise usually has surprisingly little effect on the quality of speech recognition. Where it does cause difficulties, the problem can, in most cases, be dealt with by the use of a noise-cancelling microphone or a throat microphone. Some of the factors influencing the success of speech recognition are shown in Fig. 1.4.

A fundamental weakness of spoken communication, which applies as much in direct communication between people as in computer recognition and synthesis, is its sequential and usually

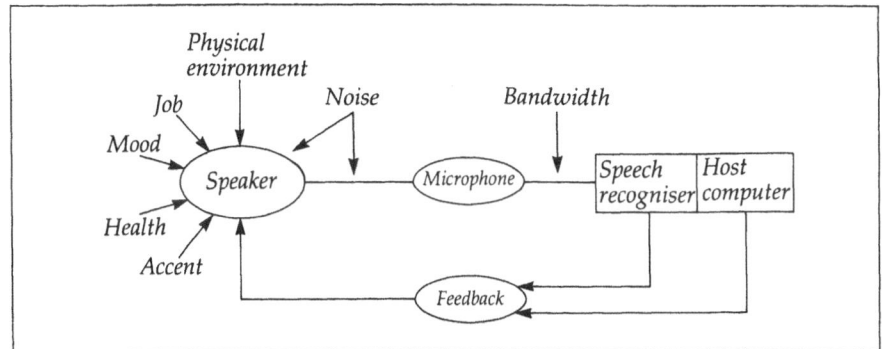

Fig. 1.4. Factors influencing the success rate in automatic speech recognition

ephemeral character. It is notoriously easy to make mistakes in hearing and remembering a lengthy spoken message. If you know that you have not heard, or have misheard part of the message, you can ask for it to be repeated, but that is time-consuming and still uncertain. There are limits to the length and complexity of messages which can usefully be synthesised if they must be understood precisely, unless they are in a highly structured form. By contrast, a written message can be skimmed, read selectively or studied in detail, and any difficult passages can be reread without difficulty.

The same applies to a lesser extent in speaking to a voice recognition system, which is comparable to dictating into a tape recorder. It is important for the system to have some means of feedback to the speaker, who may have been distracted and have lost the thread of what has been said. There are situations where voice recognition may be no faster than a written or typed message, and it may be less reliable.

These remarks are not intended to discourage the use of speech technology, but to underline the fact that it must be used with discrimination, making allowance for its limitations and taking advantage of the great benefits which it can bring.

2 RECOGNISING SPEECH

BECAUSE speech is a skill which can be largely mastered by a two-year-old infant, it is something we too easily take for granted. Speaking is a highly complex process which would take much more than this volume to describe adequately. Recognising and understanding what is spoken is even more complicated.

Producing an acceptable imitation of the human voice has its problems, as we shall see in the next chapter, but the opposite task, of recognising speech, is many times more difficult. The problems multiply rapidly as one progresses from distinguishing between a small number of distinct sounds to 'understanding' naturally spoken language using a large vocabulary.

People have been wrestling with the problems of speech recognition for more than 40 years, and millions of pounds have been spent on them by governments and private organisations. The reason is obvious – if people could talk to a computer in the same way as they talk to each other, instead of having to go through the clumsy intermediary of a keyboard, the scope for using computers in everyday life would be vastly increased.

What are the main problems?

- *Co-articulation.* When people talk naturally, they do not speak words separately with a space between them. Ifwrittenlanguagewerereproducedthesameway it would not be so easy to understand (we reproduced?), and the same difficulties arise with continuous speech. How does the computer distinguish between 'grey tape' and 'great ape', or between 'sixteen ages' and 'six teenagers', for example? But that is only the beginning. In speech the difficulties are greater because sounds are modified as the words are run together. Try saying 'bread and butter' as three separate words and you will appreciate the difference. In ordinary speech you are more likely to say 'bread'n'butter', or even 'brembutter' if you are in a hurry.
- *Different speakers.* Accents are an obvious source of variation between different speakers, but are only a small part of the problem. Even the same person's voice will vary at different times because of a cold, or because he is out of breath or worried. Between speakers there are differences because people differ in the shapes of their voice organs or the way they move them, and superimposed on these variations are differences of regional accents. A male supporter of Tottenham Hotspur on the terraces could sound quite different from a Scottish woman TV news reader.
- *Ambiguities.* In many cases, words can only be identified by their context because there are two or more words with identical pronunciations, like 'great' and 'grate'. This can happen with very commonly used words – particularly with 'to', 'too' and 'two', where a whole sentence may have to be analysed with some intelligence to establish which is the correct word. Understanding can also be hindered by our habit of using the same word with different meanings and with different functions. 'Down' may be used as a preposition, an adverb, a verb, an adjective, and as a noun with several different meanings. It is also liable to be pronounced 'darn' in London and 'dane' in Berkshire.

Because of these and other hazards, the final objective of a computer which can understand everyday speech is a very distant target. However, there are many steps on the way to that ultimate end, and a great many applications for systems offering less than total speech understanding. Some of the steps on the road could be set out as follows:

- Recognition of a fixed vocabulary of separate words or phrases spoken by one person.
- Recognition of a fixed vocabulary of separate words spoken by different people.
- Recognition of 'connected speech', in which the spaces between words are removed but each word is articulated carefully, and the overall length of the message is limited.
- Recognition of 'continuous speech', which is similar to the above but without the restriction on message length.

There are some variations in the way different people interpret the words 'connected' and 'continuous', but all of the above four steps have been taken and equipment is available which exploits these advances. There are, however, limitations on the size of vocabulary, the variety of speakers tolerated and the accuracy of recognition.

Between what has been achieved and the distant goal of 'understanding' fluent natural speech a few tentative steps are being taken, and we shall look at some of these in Chapter 8.

WHAT IS SPEECH?

Before looking in more detail at speech recognisers as we have them today, it will be helpful to review briefly the development of speech recognition over the past few decades which has led to the various systems currently in use.

Speech can be described in a number of different ways. At the most basic level it is a series of sounds produced when air is forced out of the lungs through the vocal organs. The forming of different sounds is a process which involves the whole vocal tract – shown in Fig. 2.1. Among the important elements are the vocal cords. These are folds of ligament which are normally parted but which can be closed to stop the flow of air from the lungs. They can also be made to vibrate, producing the voiced vowel sound like 'Ah'. They are important in determining the pitch of the voice, and contribute to its

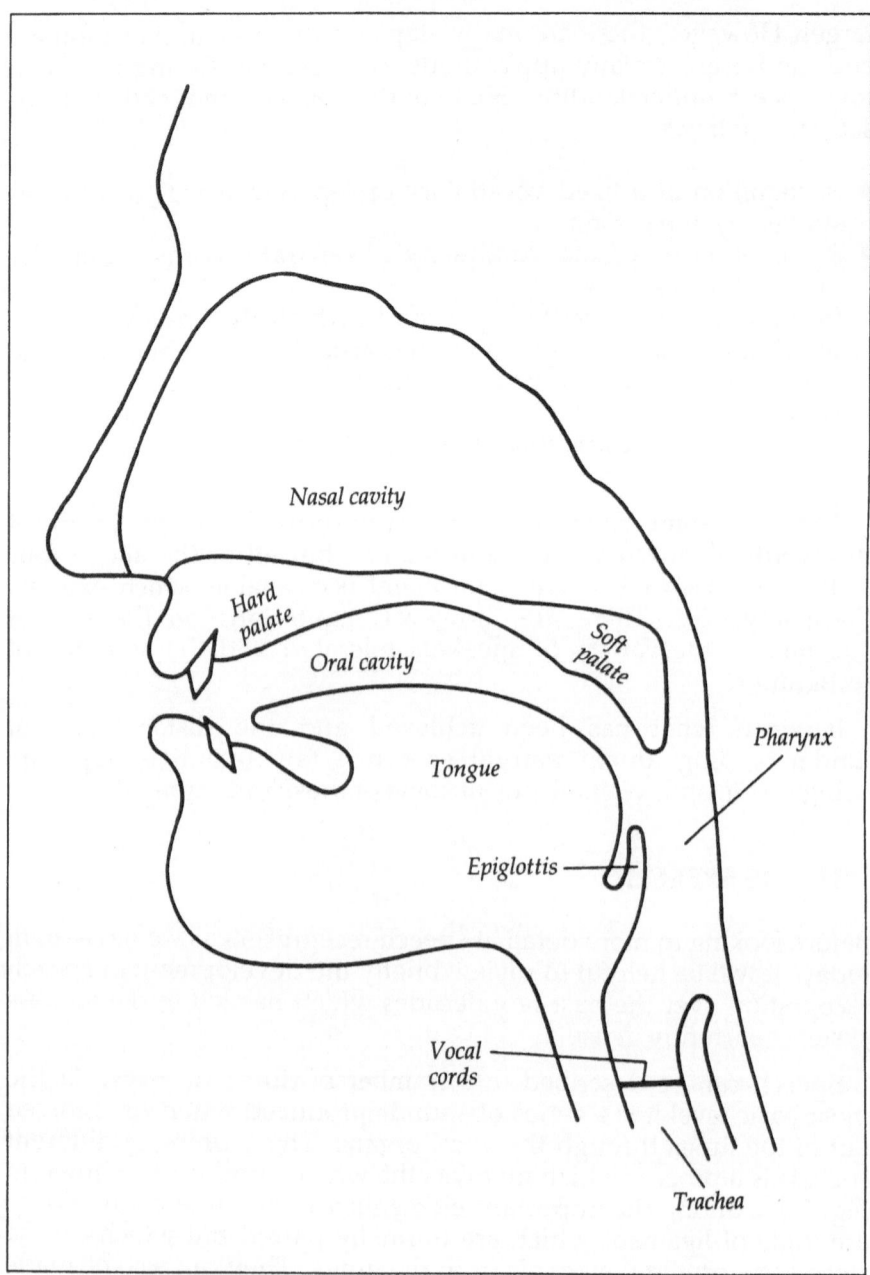

Fig. 2.1. *The sounds of speech are influenced by the whole vocal tract*

loudness and to the openness or 'thinness' of the voice produced by generating a greater proportion of high harmonic frequencies.

The cavities above the vocal cords – pharynx, mouth and nose – can be controlled to alter the resonance of the voice, and these combined with the use of the tongue, palate, teeth and lips produce the great variety of 'special effects' needed to form words in any language.

It is the task of *phoneticians* to define the different types of sounds produced in speech. They call these *phonemes*, and distinguish four broad types of consonant sounds, each of which can be voiced or unvoiced, and which can be combined in various ways. There are the sounds produced by stopping the flow of air and releasing it suddenly to produce sounds like *k, g, p* or *b*, or with a roll to make *rr* sounds. There are sounds produced by air turbulence somewhere in the vocal tract to make *s, f, z, v* and similar effects. There are so-called lateral sounds, produced by blocking the direct passage of air but allowing it to pass round the sides of the tongue, as at the beginning of the word *lateral*, and finally there are the frictionless consonants like *w* and *y*.

Early attempts at recognition involved phonetic analysis along these lines, but this method did not prove very fruitful. A different approach, which has been adopted since 1952, has led to the development of most of the speech recognition systems on the market today.

The new approach ignored the problems of analysing speech into its constituent sounds, and simply looked at the acoustic pattern produced by the speaking of a single word. This pattern of sound can be represented graphically using an instrument which splits the sound it hears into a number of narrow frequency bands and measures the volume of sound at each frequency during the time that the word is spoken.

An example of three speech spectrograms produced by one such instrument, known as a channel vocoder analyser, is shown in Fig. 2.2. Each vertical column of blobs represents, at one instant of time, the frequencies of sound being produced in an utterance, with the highest frequencies at the top. The size of each blob represents the amplitude of sound at that frequency. The change in the pattern of blobs horizontally represents the change of amplitude at each frequency level during the speaking of the word. Fig. 2.2 shows

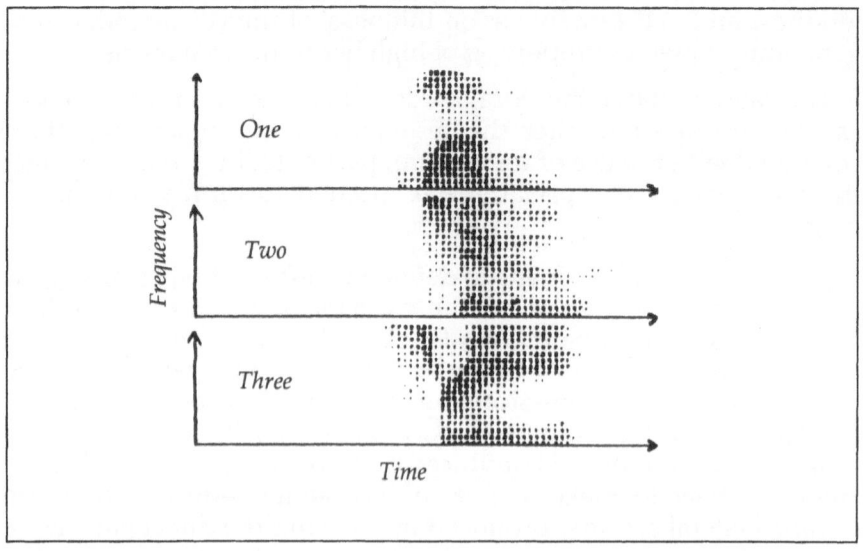

Fig. 2.2. Speech spectrograms show the different frequency distributions of the words 'one', 'two' and 'three'

quite clearly the difference between the speech spectrograms corresponding to spoken words 'one', 'two' and 'three'.

It is possible to use a set of such patterns for the 10 digits as templates with which any new utterance can be compared to find the best match, and it is developments of this technique of template matching which are used in most of today's speech recognition systems. These developments have produced vast improvements in consistency and reliability.

However, by the late 1960s some researchers had become highly dissatisfied with the inherent limitations of this mechanistic approach, and they began a change of emphasis away from engineering and acoustics. Previously the view had been that speech signals contained too much information, and the aim had been to remove the redundant material – those variations which can be heard when different people speak the same word, or even when the same person repeats the word – and to get down to what they thought were the unvarying features of speech sounds. The new line of thought was that speech signals alone would never be able to distinguish between 'through' and 'threw' or tell 'in distress' from 'in this dress'. There would have to be additional information available such as rules about grammar and the meanings of words, particularly if there was to be progress in recognising continuous speech rather than separate words.

So, in the early 1970s the emphasis of research shifted towards recognising continuous speech, and a different approach was adopted, based on the perception that people rely for understanding not only on the speech sounds themselves but on their whole knowledge of language, syntax and so on. Many spoken words are redundant and, on the other hand, people 'hear' words which are not even spoken. The goal therefore became one of not simply recognising words but of understanding the message. The criterion of success in talking to a machine was not whether the machine recognised your words but whether it did what you asked it to do.

About this time the US Advanced Research Projects Agency – ARPA – produced a report in which it proposed a new type of recogniser called a Speech Understanding System, in which recognition was divided into a number of levels each containing independent sources of speech knowledge such as acoustics, phonetics, syntax and semantics. The knowledge sources would then interact to generate an appropriate action in response to a spoken request.

The report led to the setting up of a multi-million dollar five-year project, supporting work at a number of US laboratories. It ran from 1971 to 1976, and its results were very disappointing. Systems were top-heavy in predictive power and did not pay enough attention to what was actually said. A review paper on the project in 1977 concluded that in terms of meeting the original specifications only one system had come up to the mark – and that was not part of the multi-million dollar project but the result of a one-man-year PhD thesis, and was based on a totally different approach from the ARPA work.

This system was called HARPY, and it adopted a mathematical approach to speech recognition which ignored all of the factors considered in the ARPA approach and did not even attempt to separate the words in a sentence. First, the system compiles a huge network which contains a representation of all the possible utterances which make grammatical sense within the scope of its vocabulary. This task is done once and remains fixed. Then, to recognise a particular utterance, it uses a technique known as dynamic programming to match the utterance against the entire network to find the path through the network which best explains the data. Identifying the path which gives the best fit also reveals the individual words in it.

Despite its apparently shallow approach to speech recognition, HARPY achieved all of the 12 ARPA project specifications. It could accept continuous speech from three male and two female speakers in a document retrieval task with a vocabulary of over 1000 words and an utterance recognition rate of 95%. Because it looked at complete phrases it avoided some problems of distinguishing words in continuous speech, yet a word sequence was implicit in the network structure. It also bypassed some problems of ambiguity by specifying a fixed language with a predefined grammar, and it used some straightforward procedures to deal with sounds that vary at junctions between words and with varying speed of speech.

Dynamic progamming's successful use in HARPY showed it to be a very powerful algorithm for dealing with acoustic signals. It is a technique analogous to critical path analysis of networks, allowing the optimum of a very large number of possible paths through a network to be found very quickly.

The disappointing conclusion to the ARPA project led researchers to look again at the actual content of the speech signal to discover if, after all, it did contain sufficient information to allow understanding. This changed emphasis was given an additional boost when a man at the Massachusetts Institute of Technology learned to read speech spectrograms.

So the pendulum swung back to the analysis of the actual sounds produced in speech, and it is this line of study which still dominates the development of present-day speech recognition systems.

The fact remains, however, that before we can overcome many of the ambiguities in spoken language and move closer to the idea of speech understanding, we shall have to take account of the higher levels of knowledge about speech. Research is now moving again in these larger areas, and Chapter 8 will review some of the more important lines of study.

SPEECH RECOGNISERS TODAY

All present-day speech recognition equipment is based on one or other of two mathematical models, or *algorithms*. One, as we have already said, is dynamic programming, or more technically 'Dynamic Time Warping' (DTW). The other, which has been introduced more recently, is known as 'Hidden Markov Modelling' (HMM), and is used, for example, in the systems developed by Dragon Systems in the USA and PA Technology in the UK.

DTW compares an actual speech pattern with a set of 'templates', which are previously recorded patterns for each word in the current vocabulary. It uses a mathematical method to stretch and squeeze the pattern to obtain the best fit with each of the templates available, in order to find the one which matches most closely. Fig. 2.3 shows this process of fitting an unknown word to the templates 'one', 'two' and 'three' to find the word to which it corresponds most closely. Good speaker-dependent implementations of this method generally only require the speaker to say each word once.

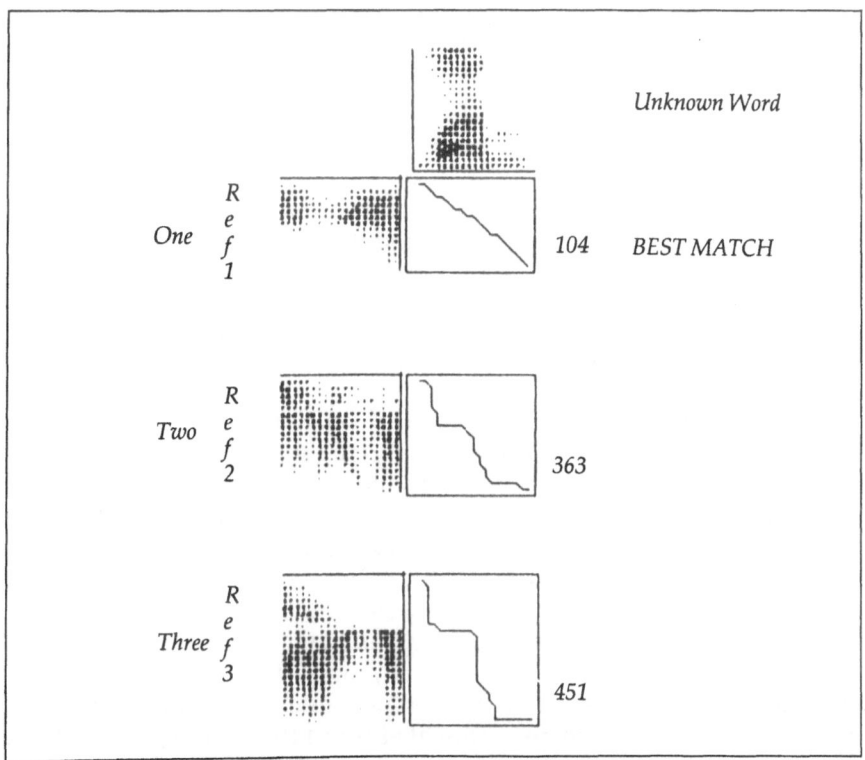

Fig. 2.3. Using dynamic time warping to find the best match between an unknown word and templates for the words 'one', 'two' and 'three'

HMM is a technique for capturing statistically the variations in the way a speaker pronounces a word. It takes a number of examples and estimates parameters of a small number of 'states'. Its underlying assumption is that speech can be described as a sequence of stationary states. A word is divided up into a fixed number of states, and the procedure calculates a series of probabilities of moving from one state to the next. Since it is a statistical method it depends on the speaker repeating each word several times in a training session.

In terms of accuracy of recognition there is little difference between good implementations of either DTW or HMM, but HMM offers some advantage in processing speed and makes possible some reduction in hardware cost in compensation for the extra time needed for training.

It is also possible to use multiple examples of an utterance with DTW. This is done by constructing a 'clustering algorithm' to find the most representative word. This technique brings DTW to practically the same level of accuracy as HMM.

Both techniques can be used to produce speaker independence. The current way most speaker-dependent recognisers are developed is to have many examples from different speakers, and then to use either clustering or Markov modelling to model the variation.

VOCABULARY SIZE

The size of the vocabulary which can be handled is important in some speaker-dependent applications, but this is a subject which is often misunderstood. As the number of words in an 'active' vocabulary increases, so too does the danger of a misrecognition. For reliable operation a recognition application should be designed so that the number of words which the recogniser can expect to hear at a particular time is as small as possible. In a DTW system this means that a spoken utterance has as few templates as possible against which it must be compared, or with HMM the various statistical probabilities are as few as possible in number. The more words there are, the greater is the likelihood that two of them will be similar, with a resultant danger of misrecognition. Great care is needed in devising active vocabularies to ensure that the words in them are as different as possible – and this needs to be checked with the individual speaker, because some words which appear to be very different produce similar speech patterns.

If a big total vocabulary is required, there must be the ability to divide it into a large number of small active vocabularies, with spoken commands to change from one vocabulary to another – some suppliers call this tree-structured arrangement of vocabularies, grammar. There should be a facility in the system's software to build such a grammar and to connect with each utterance an appropriate input to the computer.

An industrial system for a task such as inspection requires both voice recognition and some form of voice prompting, with software to control the sequence of prompts, the choice of current vocabulary and the responses to input commands. There are human as well as technical factors in the design of such a dialogue and we look at this subject in more detail in Chapter 5.

3 REPRODUCING SPEECH

THIS CHAPTER deals with the various ways in which we can use speech as an output – for voice mail, a telephone enquiry system, or in association with speech recognition.

The simplest way to reproduce a human voice is with a tape-recorder, and nothing more complicated than this is needed to run an ordinary telephone answering machine. There are several reasons, though, why this simple solution may not be adequate:

- A tape-recorder may not be a convenient interface for computer or other equipment with which the voice must be associated.
- We may have lengthy messages and need to reduce the memory space they occupy.
- We may want to compress a message or several messages so that we can place them on a memory chip using as little space as possible.

- We may want to construct a number of different messages from a stock of words or phrases held in a computer – for example to provide a speaking clock or a list of train times.
- We may want to convert lengthy, unstructured passages of text, such as a weather forecast or a whole book, into spoken output – for telephone transmission or as an aid to the blind. This will involve the conversion of text, whether held in a computer or read from printed text using optical character recognition, into continuous and acceptable speech.

The first and most important task is to put the message into a form which is as compact as possible and can easily be manipulated by a computer.

SPEECH COMPRESSION

The best quality recordings of voice and music today are digital. In very simple terms they work by sampling the sound at very frequent intervals and recording, as a number, its amplitude. The quality of the final result depends on the frequency with which samples are taken and the accuracy with which the amplitude is converted to a binary number.

One simple and perhaps surprising rule about sampling is that if there is a top limit to the frequency of the sound signal then the signal can be completely defined by taking samples at twice that frequency. The other component of recording quality, the accuracy of recording the sound amplitude of a sample, depends on the number of steps into which the whole amplitude scale is divided. In computer terms this is fixed by the number of binary digits – bits – which are allocated to the measurement. So an eight-bit scale will allow 256 divisions. The overall quality of a recorded signal is fixed by the product of these two quantities, measured as a *bit rate* of bits per second.

Top-quality microphones and recording equipment can accept sounds above the highest frequency which can be heard by the human ear, so that they can use a bit rate as high as 100 000 bits per second. This is acceptable for compact disks and tape-recordings, but is quite impractical for an industrial or telecommunications application requiring a large amount of spoken response. It would be enough to fill the random access memory of a microcomputer in a very few seconds.

Fortunately there is no need for a bit rate anything like as high as this. A telephone does not transmit frequencies higher than about 3400 Hz, so that sampling a telephone message more often than about 10 000 times a second would be a waste of effort.

The technique of sampling the amplitude at regular intervals is called Pulse Code Modulation, and it has been in use for many years. It is effective, but its disadvantage is that it occupies a lot more memory than is convenient for practical and economical applications. In consequence various refinements of the technique have been developed, with such names as Delta Modulation and Continuously Variable Slope Delta Modulation (CVSD). Such methods have made it possible to store and transmit voice messages with reasonable quality at bit rates as low as about 16 000 per second (16kbit/s) and good quality at twice that rate.

The bit rates associated with various methods of coding and synthesis are shown diagrammatically in Fig. 3.1.

SPEECH SYNTHESIS

Even bit rates of 16kbit/s are inconveniently high if speech is to be stored compactly and inexpensively on a chip, or if it is to be generated in real time by conversion from text, so the specialists have looked for other ways of bringing down still further the amount of information that has to be stored and transmitted in a speech message. They have tackled this by making use of the fact that we are dealing not with music or arbitrary noise, but with the human voice, which has certain distinctive characteristics of its own. The approach is given the name 'synthesis by analysis' to distinguish it from 'synthesis by rule' which is the method used in converting text messages into spoken output.

At present there are three main techniques in use for generating synthetic versions of human speech known as formant synthesis, linear predictive coding and time domain synthesis. Each is quite different from the others and requires different combinations of hardware and software techniques for its implementation, so if you embark on one approach you cannot easily switch to another. They differ from straightforward voice compression in that they allow a

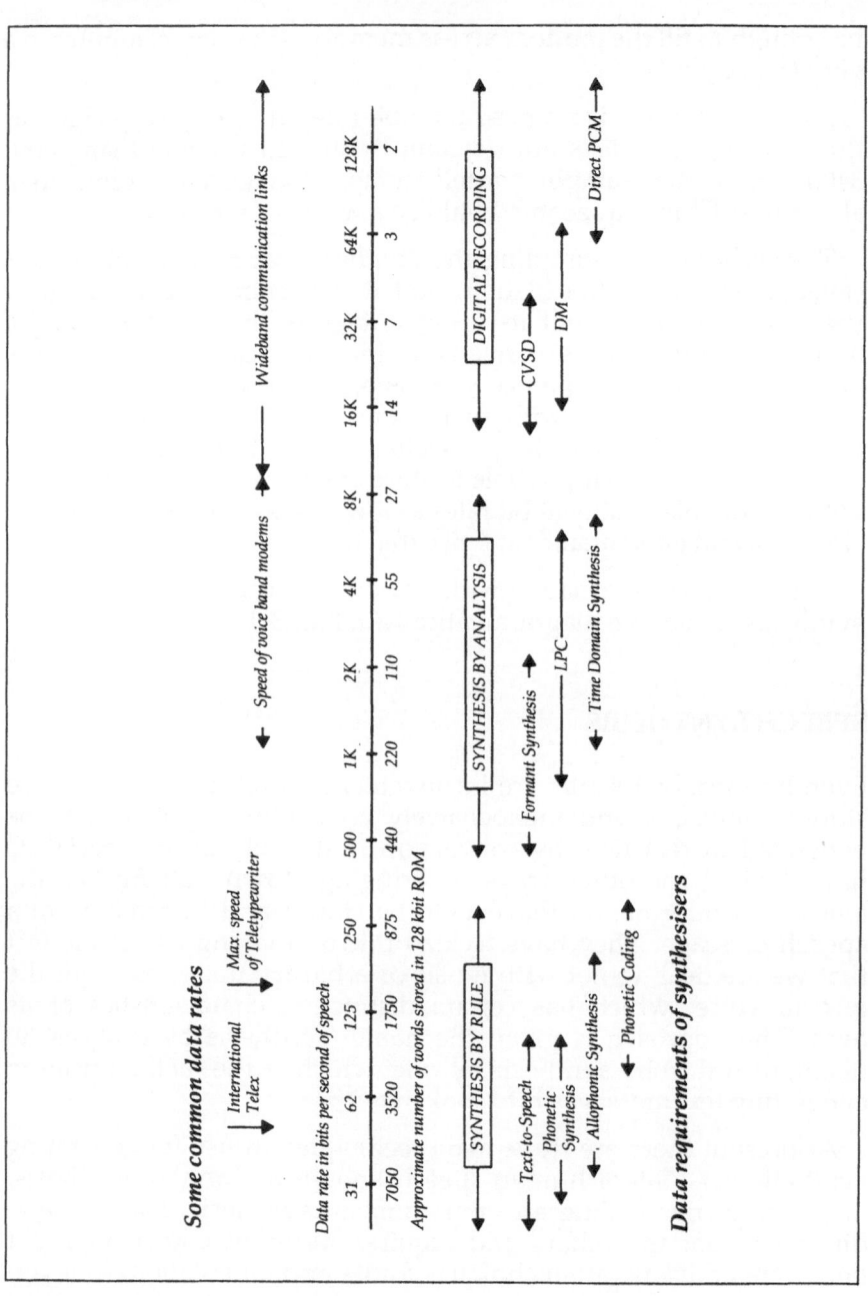

Fig. 3.1.   Data rates associated with various speech synthesis methods

variety of messages to be built up from synthesised elements. However, this requires a certain amount of additional work to tailor the speech signals into clearly understandable elements which will fit together in a way that does not sound too artificial. They start with a series of messages spoken by a 'real human being' and use their different techniques to convert the messages into highly compressed synthetic signals. If complete sentences are converted from human to synthetic speech the intonation – the natural rise and fall of the voice during speech – is converted at the same time, but if sentences must be reconstructed from word elements, some additional work must be done to blend the words together into a musical sentence – unless a monotone 'Dalek' speech is acceptable.

Formant synthesis

The human voice, like a musical instrument, gains its characteristic quality by concentrating much of the sound produced into a small number of frequency bands called *formants*. There are typically four or five such formants to be found below 5kHz in speech. These formant frequencies are most clearly distinguished in the spoken vowels and voiced consonants.

Speech synthesisers using formant synthesis build a voice from first principles, as it were, using four or five tunable resonators at different frequencies to mimic in simplified form the sounds produced in the human vocal tract. Each resonator is driven by a train of sharp pulses at a frequency corresponding to the required pitch.

Unvoiced sounds like *t*, *f* and *p*, and the corresponding elements of voiced sounds, are not formed simply at these resonant frequencies, and are modelled in the synthesiser as random noise. Other consonant sounds are generated by varying the balance between the resonators and the random noise generator. The result is a simplified modelling of speech which can, working from good original models, sound quite realistic.

Formant synthesis is an approach to modelling of speech which dates back more than 60 years and lends itself easily to the analogue methods which were employed until recently. However, it has recently reappeared in digital form and there are now several systems on the market. Fig. 3.2 shows a screen display of a Philips speech development system allowing modification of the five formant frequencies and other characteristics. Formant synthesis is

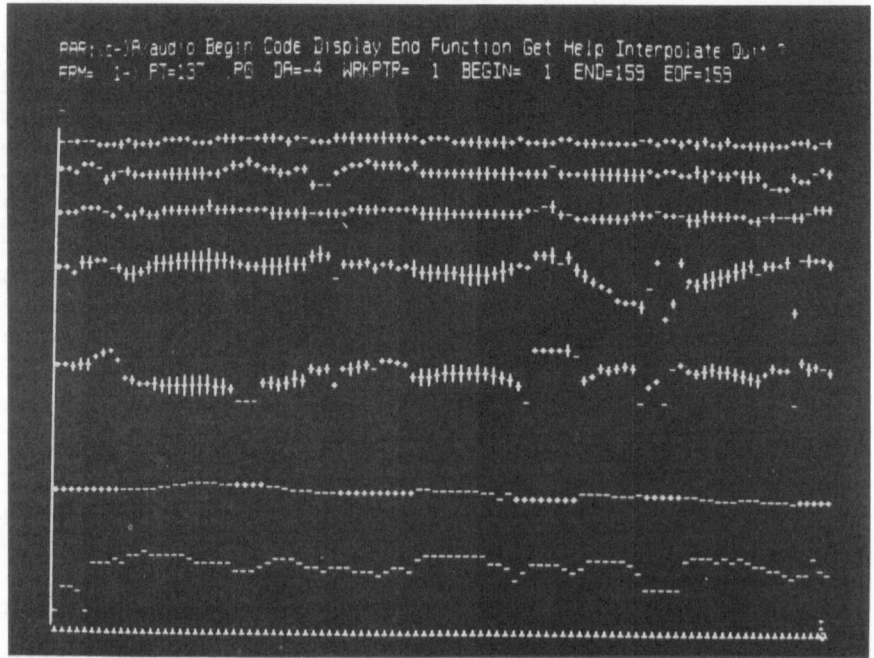

Fig. 3.2. Parameter edit mode screen display from a Philips speech development system in which pitch, contour, amplitude and rhythm are adjustable. The top five traces represent the five formant frequencies and bandwidth; the next trace represents the pitch and voiced/unvoiced source selection, the next the amplitude, and the bottom trace represents the speech frame duration

able to produce good quality speech at bit rates as low as 1–2kbit/s and reasonable quality at 500bits/s.

Linear predictive coding

The first of the modern techniques for computer synthesis of speech to reach commercial application was known as *linear prediction*, and it is still the most popularly used technique on the market. It was first applied to speech analysis in 1972.

Instead of using a number of resonators to generate the main formant frequencies of the human voice, linear predictive coding – LPC – uses a device called a multi-pole filter which can mimic not simply the four or five main formant frequencies but the entire 'spectral envelope' or cross-section of frequencies produced in a speech sound, with an accuracy which depends on the design of the synthesiser chip.

There are a number of LPC systems on the market, and the bit rate can be as low as 800bits/s, though high quality output would require 5kbits/s or higher. The technique has been incorporated in dedicated chips by several companies.

Time domain synthesis

Quite different from modelling the frequencies of human speech, used in the above methods, are the waveform encoding techniques used in the Mozer method developed by Forrest Mozer, and in the Sound Synthesis Microcomputer from NEC in Japan. These systems store compressed representations of a synthetic speech waveform as a function of time, and replay these in place of the natural speech waveforms. The synthetic waveforms are quite different, but sound correct to the ear. Performance is similar to LPC. More work is involved in data preparation but overall cost may be lower for large volumes of chips.

DEDICATED SPEECH CHIPS

A company or organisation starting to look at opportunities for using synthetic speech for standardised output of instructions, requests or information will probably begin by experimenting with an evaluation board carrying a number of standard words and phrases, to which may be added some customised phrases programmed in by the supplier to a script provided by the customer. The cost commitment at this stage is quite low, but allowance will have to be made for clumsy operation, relatively poor quality of speech output, and unnatural transitions between words and phrases if an attempt is made to run them together or 'concatenate' them. This stage, however, will make it possible to rethink the structures of messages, to experiment with different types of wording and to obtain end user reactions.

The next step will probably depend on whether the intended speech system is for high-volume incorporation into the company's products or is for low-volume or single application use with the possibility of reprogramming. Dedicated single chips or speech boards generally come from specialist manufacturers or suppliers, working from scripts or from tape-recorded material from the customer if a particular voice is required. There are many specialist skills in the practical preparation of data, their recording and editing so that a relatively realistic language flow is obtained when the various elements are concatenated.

For low-volume or experimental use, the choice will probably be a programmable speech system into which a number of preprogrammed speech chips can be plugged. Control of the whole system is carried out in software.

Just as with computers, the production of synthetic speech requires both hardware and software, and of the two the software requires the greater investment over time.

In the case of speech synthesis the software includes the messages which are stored in the system to be played back when necessary, the vocabulary of words or sounds from which the synthetic messages are to be built, and the rules and procedures for building up a synthesised message.

Unless the messages to be presented are highly standardised, however, it is probably better to avoid altogether speech synthesis by analysis, and use either compressed speech or a text to speech system. Most modern industrial inspection and similar applications and telephone voice response systems use one or both of these methods. Synthesis has the advantage of compactness and is highly suitable for incorporation into end products manufactured in volume, like motor vehicles and entertainment goods, but there is inevitably a degree of artificiality about it which can be particularly annoying to people using prompted voice input. An exception is where a very small and simple vocabulary is required, for example to speak the numerical digits.

TEXT TO SPEECH

More and more situations are arising where information in the form of computer-held text needs to be converted into speech. With the help of text to speech synthesis a person with no more than a telephone can gain access to large databases of information. Where companies have electronic mailboxes for receiving computer-transmitted text for individual box-holders, the box-holder can phone in to the office and hear a spoken transcription of any messages.

Text to speech is also useful as an alternative to compressed speech in situations where the voice message is long or has to be changed frequently, as for example in a weather forecasting service.

There is also an important market for text to speech products to assist blind and partially sighted people, where it is useful – though costly – to link it with optical character recognition equipment to allow reading of books and magazines. Some newspapers and magazines are already available in ASCII text format via electronic mail services, and so can be converted directly.

The major disadvantage of text to speech as against compressed speech and, to a lesser degree, synthesised speech is that in its present stage of development it sounds somewhat artificial. So, for example, in an industrial application where the user has to listen to the same prompts repeated over days and weeks there is usually a strong preference for good quality compressed speech.

The main hurdle to be overcome in text to speech synthesis is that written English – and any other language for that matter – is not phonetic. English in particular has many words which do not approximate even remotely to a phonetic format. Therefore, there must be at least two basic steps in creating text to speech output. The first involves converting plain English text into some intermediate phonetic language in which correct values are inserted for all the awkward words like 'one', and words are substituted for abbreviations like 'Mr' and '£'. The second step is then to use a set of rules for converting the phonetic symbols into speech sounds which can be output through the system's amplifier. To achieve anything approaching a natural speech sound there also needs to be an intermediate step – the insertion of some inflection, stress and pitch variation corresponding to the structure and punctuation of a spoken sentence. Even after all this work on the text there may be some words which are so badly mispronounced as to be unidentifiable.

There are now several different text to speech systems on the market, employing two or three different algorithms for production of speech from phonetic text. Some of them now function entirely within computer software, allowing output through any of a variety of computer equipment.

Some of the simplest and least expensive systems are based on the Votrax SC-01 phoneme-based synthesiser chip which produces

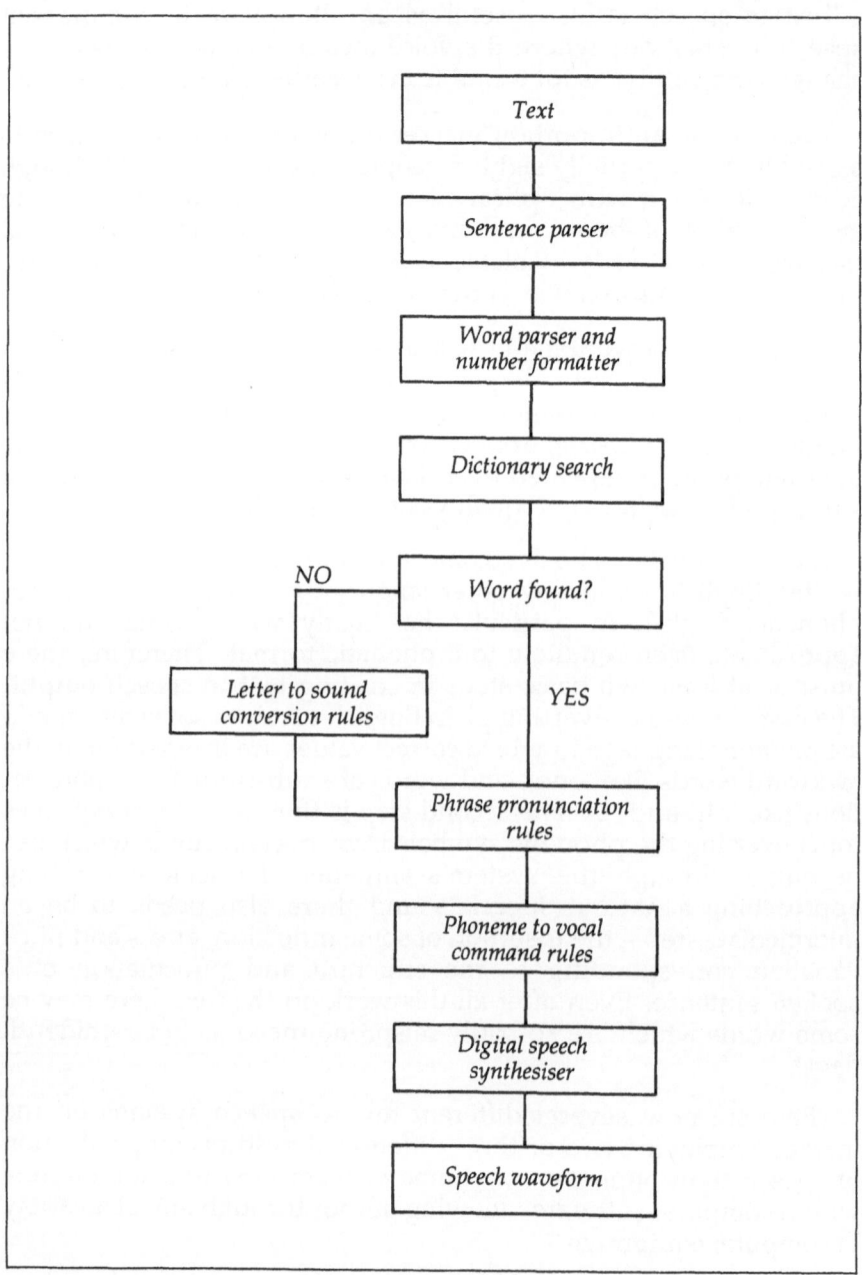

Fig. 3.3. Flow chart for the DECtalk text to speech system showing the steps in converting a computer text file into spoken language

output sounds in response to input codes corresponding to 44 phoneme sounds. It also provides pronunciation of digits, £, $ and some arithmetical functions, and has a limited inflexion capability. A practical system needs a front end for converting standard text into phonetic symbols, and some 'tweaking' of spellings can improve the quality of output.

A full text to speech system needs a more complex sequence of activities, such as that shown in Fig. 3.3, which is a flowchart for the DECtalk system. First, a sentence parser takes the ASCII text from, for example, a word processor, and divides it into separate words and user commands such as punctuation marks. Next, a word parser divides up any compound words and strips prefixes and suffixes from root words. Numbers are treated separately so that, for example, 13 can be pronounced thirteen and not one three, and similarly for larger and decimal numbers.

The system now looks up each word in a built-in dictionary covering the most common words which do not obey ordinary pronunciation rules. To this dictionary the user can add, in the case of DECtalk, up to 100 additional words such as trade names which belong to a specialist vocabulary. If the word is not found in either dictionary, the letter-to-sound module calls on the 500 linguistic rules – such as that soft 'c' precedes 'e' to generate the best estimate of the word's pronunciation. People who develop expertise in using text to speech systems learn how to introduce misspellings to correct some of the failures in pronunciation.

With DECtalk the next stop is to apply pronunciation rules to complete phrases and sentences, taking into account the transitions from one word to the next, placing of stress at the right points in the sentence, and the rise and fall of the voice from beginning to end of the sentence, punctuation marks and, in particular, the difference between a statement and a question.

The system allows the user to intervene in the otherwise automatic process and adjust a number of characteristics of the spoken output such as speed and length of pauses. It supplies nine different voices – five male and four female – as standard, and the user can also generate new voices or modify existing voices by adjusting subtle characteristics like the range over which voice pitch varies, breathiness, richness, and the size of the speaker's head!

Despite the many refinements available in this and other text to speech systems, the resultant output could not today be confused with live speech. Putting more life and realism into speech synthesised by rule will require something like the application of artificial intelligence to the content of the text, and work along these lines is in progress.

4 THE SCOPE FOR SPEECH TECHNOLOGY

PART of the difficulty in achieving the wider use of speech technology is simply a failure to realise what its possibilities are. Equally there are many applications one can envisage which, for technical or economic reasons, are not currently worth pursuing.

This chapter outlines a number of reported applications of speech technology across a wide spectrum of industry and commerce. We make no claims for their economic success and report them simply to set you thinking about possible uses for speech technology. Wherever possible the source of the information is quoted. Chapter 7 gives some more detailed information on selected applications which have been fully documented.

Applications of speech technology to date have advanced mainly in a few well-defined areas:

- Storing and reproducing compressed messages, generally known as voice messaging.
- Telephone systems using synthesised voice response.
- Industrial systems using voice input and response.
- Military applications using voice input and response.
- Aids to the handicapped.
- Educational aids.
- Consumer products.
- Speech to text systems, though not the long-promised 'voice typewriter', have arrived for strictly limited applications.

VOICE MESSAGING

Voice messaging is a system for compressing and storing spoken telephone messages for retransmission later, making it a voice equivalent of electronic mail. There are a number of such facilities offered as a bureau service in the USA and Europe, a British version being British Telecom's Voicebank, which provides a 24-hour seven-days-a-week service. Many major companies have voice messaging systems of their own which allow incoming messages to be stored if the recipient is not at his or her desk, and transmitted over the telephone later, after the style of a telephone answering machine. Most systems currently store messages with a density of 18 to 32kbits per second – 24kbits/s will fill a 10Mbyte Winchester disk in a little under an hour's recording. Ferranti's Voice Manager system, for example, is available in versions supporting 100 or 600 voice mailboxes. Other systems can offer 3000 or more mailboxes.

At its simplest, voice messaging is no more than a digital equivalent of the telephone answering machine, but all commercial systems offer more than simply storing and forwarding messages. They are usually integrated with a company's PABX (Private Automatic Branch Exchange), and can intercept calls, send messages to callers and store callers' messages with separate mailboxes for all recipients. Each message can be time and date stamped, and given a priority rating. All the mailboxes can be given security facilities to prevent unauthorised access to the messages, but the recipient can call from outside to pick up any waiting messages. There can be links to paging systems, and systems can be integrated with voice response and even voice recognition systems which identify the caller.

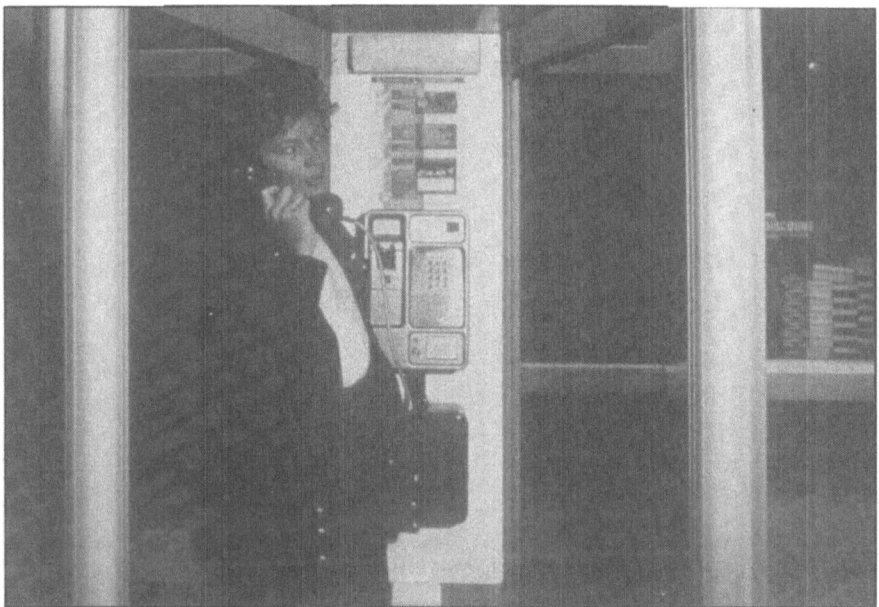

Fig. 4.1. Voice database enquiry systems allow access to a database with no more equipment than a telephone

There are many ways in which voice messaging can be used. For example, the caller can be given a message offering the choice of leaving a mailbox message or of speaking to the person being called. There can also be a facility for speaking to the operator. The system is not confined to receiving incoming telephone calls. The facility for recording messages simultaneously in any number of mailboxes can be used to run a company voice-based bulletin board. Some companies find it a useful alternative to the internal memorandum. Some systems allow the sender to check automatically whether the message has actually been played back by the recipient – which is more than can be done with a memo.

Major users of voice messaging find that the initial dislike of impersonal responses, which is encountered even by users of telephone answering machines, is quickly overcome. The international company 3M has been using voice messaging since 1981, and now has 8000 of its 90 000 employees worldwide connected to a voice system from VMX. The comany has found that voice messages left in mailboxes are getting longer and takes this to be an indication of growing acceptance. Instead of simply leaving a request to call back, users are leaving lengthy messages in the mailbox, and often get lengthy replies delivered to their own mailboxes, so that people are talking to each other via their mailboxes.

TELEPHONE VOICE RESPONSE: KEYPAD SYSTEMS

The use of DTMF keypad dialling in the American telephone network allows the user to send digital information over the telephone line once the call has been connected. This has made possible the growth of a great variety of automated information services and other facilities based on synthesised and compressed voice response. They include added sophistication in voice messaging, access to databases and information services, and can include input of digital information as well as voice response.

The pulse dialling method which until recently was the only one in use on the British telephone network has restricted the growth of such applications in the UK, though it is possible to connect a separate tone keypad to the telephone for sending information.

Small business facility

Facilities do not need to be highly sophisticated to take advantage of tone dialling. For example Natural Microsystems in the USA offers a 'Voice Branch Exchange' – VBX – based on an IBM PC which is suitable for small businesses with up to 100 employees. It automatically answers incoming calls and routes them to the proper extensions. It records and delivers voice messages. It can be used to supply product and service information, and can even carry out telephone marketing and market research activities automatically – all with the help of the DTMF keypad, which is at least as important an element in many such systems as is the voice compression and synthesis.

With the VBX, callers hear a message offering the options. If callers know the extension number they want they can simply dial it. If not, they can dial for a spoken list of the company's employees and extension numbers in alphabetical order. Just before putting through a call, the VBX asks for the caller's name, and uses it to announce the call. The recipient has the option of taking the call or asking that a message be taken. This activates the voice mail facility, which can hold up to 250 voice mailboxes.

The telephone marketing facility will automatically call a series of telephone numbers and deliver a message to each person who answers. It can provide the opportunity for some degree of interaction by asking questions which can be answered by entering numbers on the DTMF tone keypad.

Postage by phone

Pitney Bowes, the international postage franking machine manufacturer, is using a voice response system to run its 'Postage by phone' system. This allows customers to buy postage and recredit their business mail systems in 90 seconds using the telephone – instead of the old system of taking the machines to the Post Office and paying cash. The machines have an electronic combination lock which requires a new combination to open the machine for resetting each time it is opened. Customers telephone the Pitney Bowes computer and follow a prescribed identification routine using a multi-frequency telephone or a separate tone keypad. They are then prompted by automatic voice response through the steps necessary to reset their franking machines. Customers hold a credit balance in an independent trust fund and their accounts are debited automatically to the Post Office by the amount of postage credited. The system also provides a facility for customers to check their credit balances. In the UK the voice response equipment used is TalkBack, supplied by Autophon.

Weather forecasting

Telephone input/output systems giving access to databases and other facilities are in use in the USA with the input side operated by the DTMF keypad. A weather forecast service for aircraft pilots is one of the more imaginative applications which was developed by IOCS in the USA in collaboration with the Federal Aeronautics Administration over a number of years since 1975 and an Interim Voice Response System came into use in 1985 following a $20 million contract with IOCS. It provides continually updated weather forecasts covering all the busiest air traffic routes in the USA on a 24-hour basis, seven days a week – all automatically.

The system is shown diagrammatically in Fig. 4.2. The user can call one of many 'remote speaking systems' in the USA, all of which are connected to a central site system at the IOCS headquarters. Here synthesised reports from the National Weather Service's Weather Message Switching Center are stored and continually

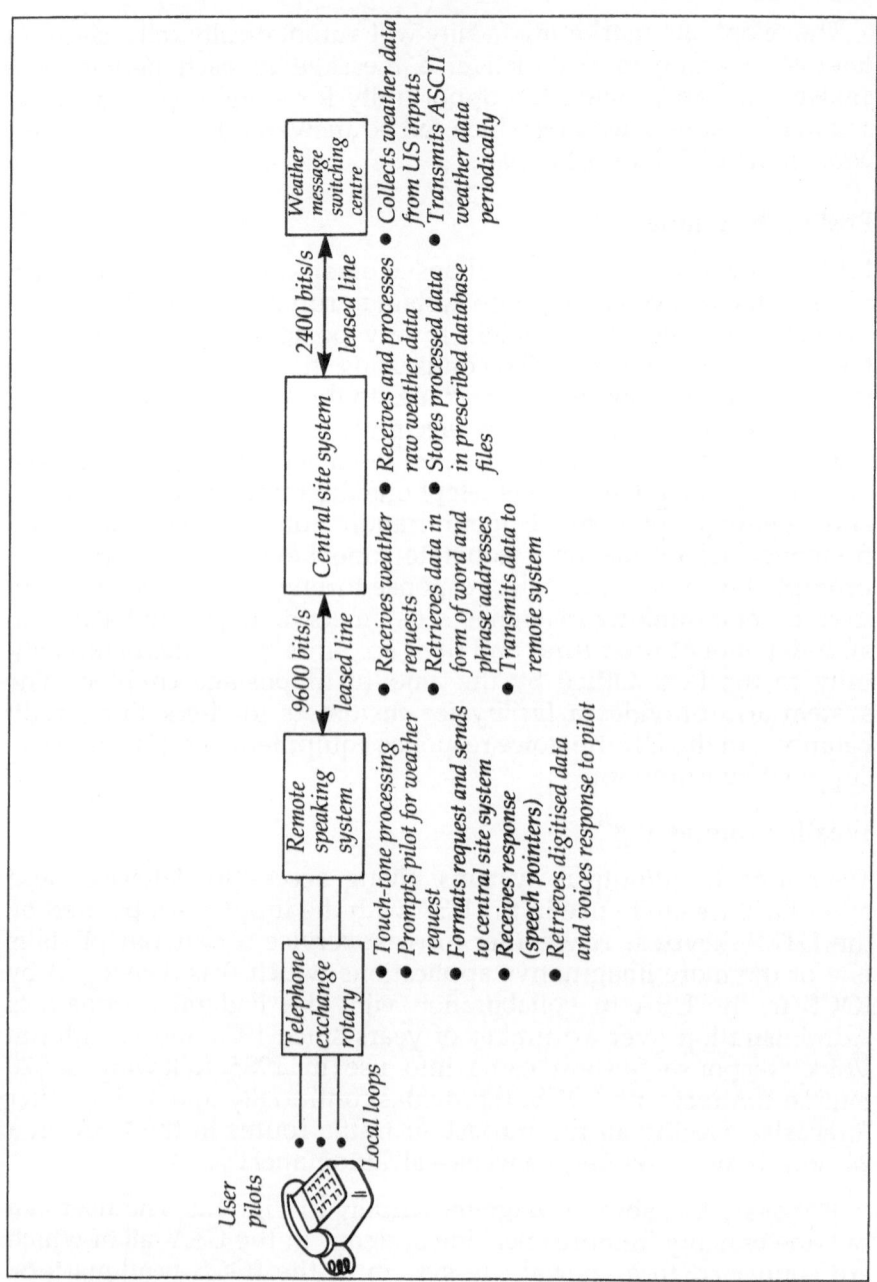

Fig. 4.2. *Voice response system in use in the USA for supplying weather forecast information to airline pilots*

updated. The user can have up to 15 minutes of access time, during which he can use the telephone keypad to enter three-character identifiers for the locations covered by weather reports, and for the route weather information given in 'Transcribed Weather Enroute Broadcasts' – TWEBs. Information supplied covers hourly surface observations, terminal forecasts, winds and temperatures aloft, convective significant meteorological conditions, severe weather forecast alerts, as well as the TWEBs.

Text to speech implementation

Where a voice response system involves lengthy messages which must be updated frequently, one approach is to use a text to speech system such as DECtalk, 'in which a word processed text message in a computer is interpreted and spoken by the text to speech system. The quality of the spoken voice is not as good as an ordinary digitised and compressed spoken message, but text to speech is more flexible in use.

A typical DECtalk implementation is run by Inlander-Steindler Paper – I-S-P – a wholesale distributor based in Illinois. The company operates six independent warehouse facilities and sales offices throughout the US midwest, and these are networked on dedicated telephone lines to a VAX 11/785 computer. Customers and sales people gain entry to the system through authorisation codes entered via the telephone keypad. Depending on the code classification, DECtalk directs the caller either to the customer order entry menu or to an extended menu available to the company's own sales force.

Once logged into the customer order entry system, the speech system asks for a purchase order number, the item number and quantity. These are confirmed verbally by DECtalk before an invoice number is issued. For example it would say 'You have ordered 10 cases of item number 2605. Description is Scott 150 Towels'. Letters as well as numbers can be entered by the use of a conversion table, which also allows the user to backtrack, repeat, cancel entries and so on. Finally the caller can ask DECtalk to read back the entire order details for verification. Another service to customers gives on-line sales history information. Buyers can phone DECtalk to find, for instance, how many towels and tissues they bought during a specified period, and they will get a spoken answer.

Facilities for the company's sales force are an expanded version of the customer order system, giving out-of-stock messages, credit authorisations, the ability to enter specific pricing and discount schedules, and other information.

Text to speech can also provide a link between electronic mail services and voice messaging. The holder of an electronic mailbox can call in to the office from an outside telephone, and can have messages converted from text to speech and read aloud – avoiding the need to carry around a portable computer with modem and acoustic coupler.

Speech paging in medicine

At Pacific Presbyterian Medical Centre in San Francisco, Dr Reid Rubsamen has been working with Berkeley Speech Technologies to develop a synthesised speech paging system on a personal computer linked to computer equipment monitoring the condition of patients in intensive care. Berkeley's text to speech equipment on the personal computer is connected to a Siemens Sirecust 454 central station patient monitor display cluster, which signals any abnormality in a patient's blood pressure, heart rate or other condition. Normally such equipment is set to sound an alarm if there is a change in the patient's condition, but this does not identify the specific cause. With Rubsamen's system, his personal computer dials the hospital paging system which sends out a call for the appropriate doctor, and when the doctor responds it transmits a text to speech message reporting the patient's condition – for example, 'Jones blood pressure systolic less than 90 for three minutes'. With such information the doctor is immediately able to telephone instructions on action that must be taken.

Time and attendance reporting

Time and attendance reporting by telephone is being used by a regional health centre near Dallas, Texas, using a SofTalk II system from Denniston and Denniston. When employees arrive at their workstations, leave or return during the day, or move to work in another department, they dial an extension within the hospital on a telephone system secure from outside access. The calls are automatically answered, and the caller is prompted through a series of spoken messages to supply employee number and transaction type – clocking in or out, changing departments and so on, by using the telephone keypad. Prompts can be given in different languages, called up automatically when the caller enters an employee number. The system automatically adds the time and date and stores each completed transaction.

The time transactions are passed daily to the hospital's host computer, which processes them automatically into a series of management reports for department supervisors and passes them to the payroll system.

Sales services

Database access applications can take many forms and can be operated in a variety of ways. An attractive sales service for some companies is to provide details of products on a free telephone line. A books retail chain could maintain lists of books that are out of print but available at certain stores, and the system could accept an order against a credit card number. Many financial and industrial databases currently available on-line to computer users can be made accessible to anybody with a telephone by the use of text to speech or concatenated synthesised output responding to DTMF input.

Stockmarket information

A telephone voice response system recently introduced in the UK, with the prospect of upgrading also to voice input in the future, is the Financial Times Cityline service, which has been implemented using standard Marconi Keycall voice response computers and specially programmed stock data processors.

The Financial Times started its own telephone service for investors when British Telecom decided to withdraw its Guideline service, which had been running for 12 years with data supplied by the Financial Times. Besides greatly expanding the amount of information available to all callers, the Financial Times decided to add a personal portfolio service available on subscription. Callers pay for the general information through British Telecom's 'Callstream' service, charged at 38p per minute during peak and standard rate times and 25p per minute during cheap rate periods from anywhere in the UK.

Like many services in the USA, the Cityline service is currently controlled by the caller with a DTMF telephone or any other telephone together with a multi-frequency tone keypad which sends the correct tones down the telephone lines when its keys are pressed.

When the caller first dials the Cityline number, the first response is the up-to-the minute FTSE 100 Index. The caller is then invited to choose another service. Entering '99' will give a list of all the financial reports available with a two-digit reference number, and entering the appropriate number will produce the desired report. For callers without a DTMF telephone or a tone-pad, a separate telephone number is available for each report.

A four-figure code, selected from a printed list, gives the current price of any of more than 3500 shares. All of these services are paid for simply by the line charge, but an additional subscription service is available which allows the caller to store a portfolio of 20 shares in the Cityline computer: on entering a personal number the system will speak the current prices of all the shares in the portfolio. As part of the same service, the caller can also request the current value of the total portfolio.

Behind this complex facility is a bank of Marconi Keycall voice response computers networked together and to the Stock Exchange database computers, with two leased lines to the International Stock Exchange SEAQ computer, which provides real-time price information on all UK alpha, beta and gamma equities and major international equities. Each Keycall has six four-channel voice cards in an IBM PC-AT compatible computer, giving 24 simultaneous telephone lines per Keycall.

Digitised natural speech is used for the voice response, and is stored on high-capacity disks giving a capacity of 26 hours of recorded speech. Messages are stored with reference numbers and short messages can be concatenated to generate complete voice responses. The FTSE 100 index, which is continuously updated, is concatenated from individual words and short phrases to produce a message like 'The FTSE 100 index is up twenty at two thousand'. The stockmarket reports are updated at regular intervals and are spoken by the Cityline operational staff and tape-recorded. The tape-recordings are digitised by one of the Keycall units and stored on disk.

Bank automation

A voice-operated extension of a bank automation system is currently (1988) being installed by Bank of Boston in the USA in collaboration with IOCS. During the past few years, the bank has been implementing a number of computer automated services, including a money and wire transfer service and a balance reporting service, but these required the customer to have his own terminal. The services are now available over the telephone with DTMF telephone keypad entry and voice response. Besides services like funds transfer the intention is also to supply information on investment rates and even to allow trading over the automated system.

TELEPHONE VOICE RESPONSE: VOICE INPUT

The above applications all depend on the availability of a DTMF keypad. A very attractive alternative would be a system which could respond to spoken input. This immediately multiplies the difficulties many times, because to be of any real use the system requires not only speech recognition but recognition of any speaker's voice on a telephone line.

Systems are in use which operate using voice input, but usually they need to be speaker-independent, which means they must use a very restricted vocabulary consisting of the numerical digits plus a few necessary words – a sort of spoken keypad.

Telephone dialling system

A relatively simple application which is not restricted by the need for speaker independence is voice-operated telephone dialling systems for cellular telephones. Examples are the Topaz system by British Telecom and the Vocalink system marketed by Interstate Voice Products in the USA.

Organ transplant bank

Using speaker-dependent input in a very interesting application is an organ transplant bank operated in the USA by the North American Transplant Coordinators' Organisation (NATCO), using a Votan Voice Management System and NATCO software. It allows rapid and efficient matching of extrarenal organs (organs other than kidneys) for patients awaiting transplants at more than 33 heart, liver and lung transplant centres. The bank matches prospective recipients with prospective post-mortem donors at hospitals across the USA and Canada. Conventional computer matching is available, but the voice system is claimed to save time and to simplify the matching of suitable organs. The system operates unattended and is not restricted to DTMF keypad telephones, which are not standard in many parts of the USA.

Before the present system came into use, a doctor was able to call a number and hear a lengthy tape-recording listing transplant organs that were required. With the new system the doctor can name the available organ and hear a much shorter list including essential requirements such as blood group in each case.

The system first checks whether the caller has a DTMF telephone by asking for a number 1 to be entered. If there is no tone signal it switches to speaker-dependent recognition, going through a brief training routine, asking the caller to speak the digits one to nine, 'yes' and 'no'. The remaining 'conversation' is carried on entirely with this vocabulary or with the keypad.

The caller is first asked in which region he wishes to search for recipients – east coast (1), west coast (2) or anywhere in the USA and Canada (3). Six different responses are offered for reporting which organs are available, and four for blood types A, B, AB and O. At this point the system plays back the responses so far and asks for confirmation. Next it asks for donor weight, age and sex, and then plays back these responses for confirmation.

At this stage the caller is asked to hold while the system switches to another program which searches for a matching recipient. It then reports back to the caller brief details of the patient in most urgent need. The caller can ask for more information or can request details of the next most urgent case. During this stage of the exchange the system holds in store the telephone numbers of the centres reported on, and at the end of the call it offers to play back to the caller the list of phone numbers. This is a simplified description of a highly sophisticated system which has become a major resource in the USA since it went on-line in 1983.

Stock quotation service

A stock quotation service for its customers is being run by Fidelity Brokerage Services of Boston in the USA using AT&T's Conversant 1 voice system. The voice system acts as a front-end processor to the company's computer system which provides quotations on demand for 6000 stocks, stock option quotations, a personal list of stocks to watch, and the current Dow Jones Industrial Average. The system can operate with a very restricted voice input vocabulary. The 6000 stocks, for example, are identified by nine-digit codes in a catalogue held by the user, allowing voice or DTMF input.

Other actual or potential applications for systems of this type include:

- Home shopping, associated with TV commercials.
- College registration for new courses.
- Remote banking, with voice or DTMF input of the user's account number and personal ID, and a spoken menu of options.

INDUSTRIAL APPLICATIONS

Applications of voice input/output in manufacturing industry fall mainly into two categories:

- Product inspection.
- Storage and warehousing.

There also appears to be scope for voice input in association with some complex manual tasks such as controlling a CAD system.

Inspection

Inspection is very often highly suitable for speech recognition because the inspector may need to travel around the factory, in some cases climbing over structures, to use hands and eyes to operate measurement and test equipment, and at the same time to make accurate notes. It may also be important for any errors or defects to be reported as quickly as possible to prevent further defective work.

Some inspection tasks require very close attention, which is distracted if notes have to be written down. In quality assurance of semiconductor wafers, for example, operators have to use high-powered microscopes to search for defects, and have to report on possible defects and their causes.

The Californian company Kevex Corp. has simplified the use of its X-ray microanalysis equipment for finding the elemental composition of samples, by integrating with it a Votan V5100 board for speech recognition and response. This allows the user to control this very complex piece of equipment while keeping his eyes focused on the specimen under investigation. The voice response facility gives spoken confirmation of his instructions and other information.

The American aircraft engine manufacturer General Electric is using speech recognition and synthesis, together with bar-code reading, to aid inspection at its Lynn, Massachusetts, factory where engines for helicopters and aircraft are built. Inspectors have to carry out detailed physical examinations including internal inspections with fibre optic borescopes. The work involves climbing over the engine, and inspectors have to review the paperwork on each engine for accuracy and completeness. They use a mobile Digital 100 computer terminal, a DECtalk speech synthesiser, an Interstate VRT 300 speech recognition board and an Intermec Code 39 bar-code reader and wand. Spoken prompts are provided by the synthesiser through the inspector's headset or a loudspeaker and he speaks his responses into a microphone. A radio link to the terminal avoids trailing wires.

The inspector's responses to the synthesised prompts follow a vocabulary which he has previously spoken to 'train' the system, and which is stored against his name. The words must be spoken separately, not connected, and form a single 200-word vocabulary for each inspector. The voice recognition board matches each word spoken against its vocabulary and sends to the synthesiser a digital code corresponding to the word it has recognised. The synthesiser then plays back to the inspector the word corresponding to the code, allowing the inspector to make a correction if necessary. Results of the inspection go automatically into a database held in a minicomputer in the quality office and can be printed out at any time.

Inspection does not follow a fixed procedure but is controlled by the inspector. Prompts from the synthesiser ask for the main assembly being inspected, to which the response could be 'fan'; then for the part, which might be 'blade'; followed by the nature of the fault – 'bent'; and the quantity – 'one'.

Supporting the voice input, more precise identification can be provided by the bar-code reader. The inspector can consult a work sheet of several hundred engine parts and manufacturing irregularities, each with a bar-code reference. He can simply wipe his wand across the appropriate codes to send to the computer a precise description of the part and the fault. There is also a portable keyboard, but most inspectors only use it to add various comments at the end of an inspection.

The DECtalk synthesiser, working as it does from computer-stored text, is not confined to single words or even to a limited vocabulary. Anything the computer terminal can display can also be spoken by the synthesiser, and appropriate punctuation can ensure, for example, that questions are spoken with a rising intonation at the end of the sentence.

Benefits from the system are dramatic, according to the company, GE. Because the inspector's findings are entered immediately into the computer there is no delay in sending reports. The information is sorted and compiled into working reports very quickly.

In another branch of GE in the USA, the Aerospace Control Systems Department, a voice recognition system is being used – again in association with bar coding – for data collection on the performance of printed circuit board assemblies used in avionics computers. The inspection situation is typical of those where speech technology is well worth investigating. The printed circuit boards have to be inspected visually on both sides. They each carry about 220 components and 2800 soldered joints. They have to be held in both hands for inspection, and about 20% of the inspection work requires the use of a low-powered microscope. Before the new system was introduced, any defect had to be marked on the board and the nature of the defect entered on a quality rejection document which travelled with the board to the rework station and then back for re-inspection. Further lengthy procedures were then initiated with the quality document itself for information purposes.

The department developed its own new system software for running four workstations using bar coding for accurate identification of products and optional voice or keyboard entry for inspection reporting. It was envisaged that the keyboard would be used to enter occasional defect codes which were not part of the voice recogniser's vocabulary, to deal with any recognition problems that might arise, and for general data entry at the discretion of the operator.

Experience with the new system was good, particularly in the quality of information going into the quality database which improved from five errors to fewer than one error per hundred forms. Errors of transpositions and incomplete product identification and document references were virtually eliminated by the use of bar codes and check digits on numeric strings.

Glass faceplates for television tubes are inspected at Owens-Illinois with the help of an NEC voice input system. Consistently high quality is demanded by customers and tolerances on the faceplates must be monitored continuously with real-time feedback to manufacturing. Voice input of inspection data allows operators to carry out verification on-line, to move about the test area and to work without having to write or key-in data. The company particularly stresses the fact that the training required is minimal. Operators do not have to memorise special codes, punch keys or record information by hand for subsequent computer entry. The system is highly flexible. It can allow for various different sample inspection plans, adjust maximum and minimum tolerance conditions and so on, and several different types of reports can be produced.

Inspection includes weight and several dimensional measurements, and all the voice reporting is done with a small speaker-dependent vocabulary of 14 words – the 10 digits and 'override', 'cancel', 'complete' and 'minus', which can be used in connected speech mode by including some connected digit strings in training the recogniser.

Other industrial situations where speech is attracting attention is where extreme cleanliness is required for microscopic assembly or inspection. The operator may have eyes and hands busy on a delicate task, and there is the added need for extreme cleanliness and avoidance of dust, which is less easily maintained with pencils and report sheets than with a speech headset.

Warehousing

Warehousing has been the subject of a number of effective applications of voice input and response. It is a situation where people need to move about, with a high content of manual work, but also with much writing and updating of records, so that there is a conflict between the needs of the manual and clerical functions.

At the US distribution centre of 3M at St Paul, Minnesota, a speech system for Interstate Voice Products is an integral part of a new Reship Order Control System – ROCS – which has greatly reduced the reshipment times of goods coming into the centre as well as improving the accuracy of shipment. Previously an in-efficient system of paperwork, coupled with the lack of effective data collection, caused delays of typically two to eight hours in reporting and routing information to its proper destination, no matter how hard employees worked.

An on-line computer system with traditional video terminals would not be adequate because the warehouse personnel would find it difficult to enter data on incoming shipments while moving boxes and pallets from incoming unloading areas to staging areas for reshipment. In addition, errors made in such manual data entry could at worst result in misplacement or loss of a load or pallet information for hours, if not days.

The solution adopted included the use of bar-code labelling of all material coming into the centre. The labels were generated on the unloading dock with a dot-matrix printer and affixed to each part of the shipment for later use for tracking and routing.

The other essential ingredient was a voice recognition system allowing power truck and fork-lift truck operators and other personnel to input data directly to the computer system.

When a trailer arrives at the dock, pallets are unloaded from it and placed in a receiving area. At that point an operator reads the information on the paperwork – invoice number, origin of the load, destination, number of pallets and so on – and speaks the information into a headset microphone, using a 54-word vocabulary which includes all alpha and numeric characters, 'yes', 'no', 'dash' and control words such as 'enter', 'send', 'backspace' and 'verify'.

The computer uses synthesised speech to confirm each item through the operator's earphones, and enter the data into its memory. Then the computer activates a printer on the warehouse floor to prepare the number of pallet labels needed to get the shipment on its way. Besides the necessary human-readable information, the label also has two identical bar codes, one of which the loader tears off and fixes to the shipping documents. At the time of pallet loading, the operator scans the label on the load, updating the computer on its status. The shipment then goes to the staging area for consolidation and final shipment.

The shipping document with the trailer identification number goes to a receiving clerk, who scans the bar code with a laser unit connected to the Distribution Centre computer. This enables the computer to pass all of the information into its own storage and to record the transaction as complete.

According to the manufacturing services data-processing supervisor, the system is very accurate. In handling materials and shipments it is achieving an accuracy rate of 99.9%. It is also saving a great amount of processing time because it is so fast.

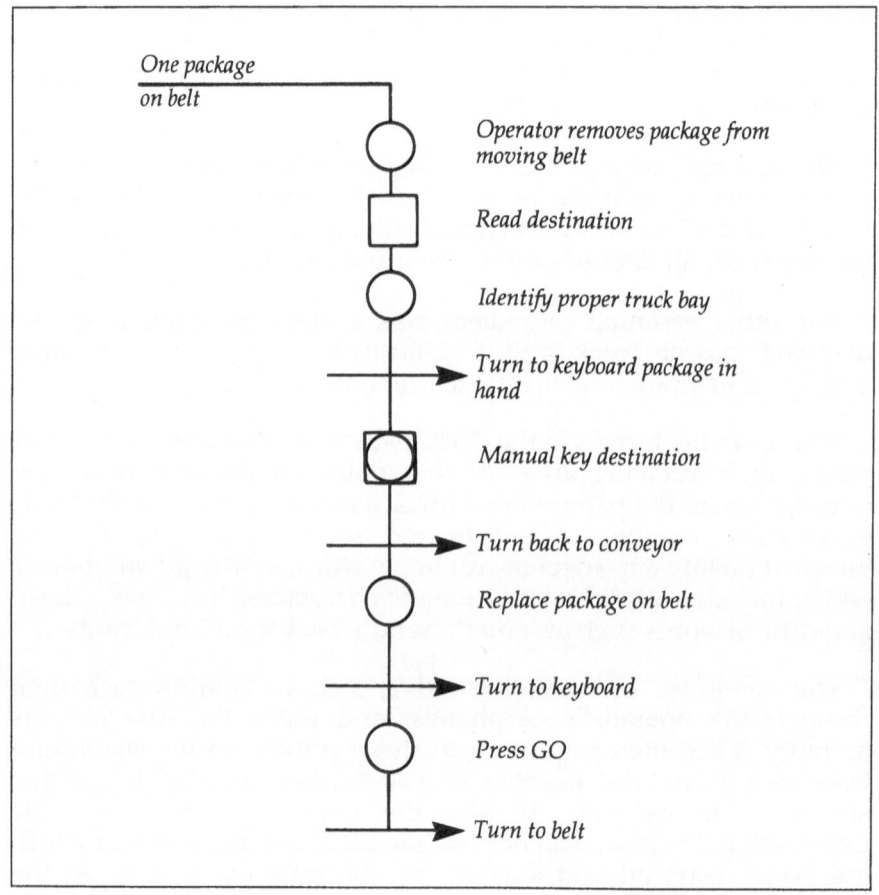

Fig. 4.3. Flowchart for a package handling application before introduction of speech input

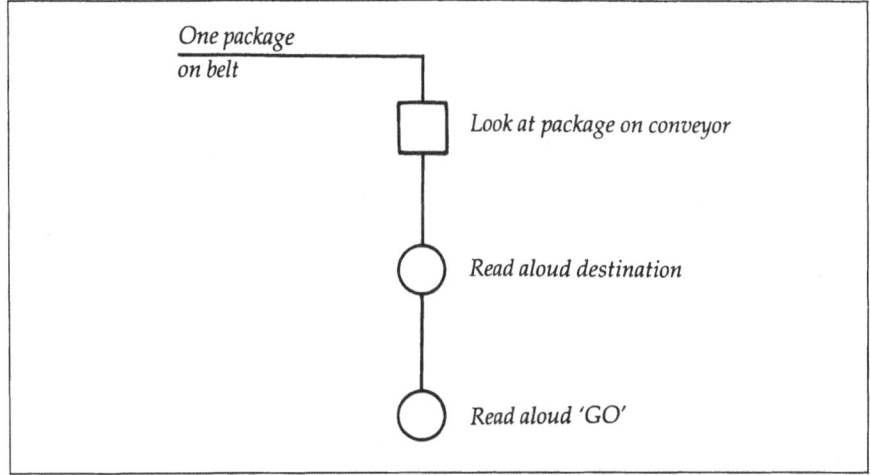

Fig. 4.4. Flowchart for a package handling application with the implementation of speech input

Package sorting and baggage handling give very good opportunities for benefiting from voice input. Typical flowcharts for a package sorting application before and after the introduction of speech are shown in Figs. 4.3 and 4.4.

United Airlines in the USA has been operating since 1984 a voice system which enables baggage handlers to input flight destinations by voice rather than by manually keying-in the information at a terminal. The company estimates that voice data entry has increased baggage handling throughput by more than 100%. With manual data entry, handlers can process about 12 bags per minute. With the Verbex Model 3000 equipment they can deal with about 30 bags a minute. According to the airline, the system enables them to enter city names of several syllables without pausing or speaking in an artificial way, in a situation of high background noise.

SURGICAL APPLICATIONS

Similar in many respects to the problems of industrial inspection are the needs of surgeons in the operating theatre. Fig. 4.5 shows a Speech Design system in use with a Carl Zeiss microscope for micro-surgery, in which spoken instructions are used to control the focus, zoom and movements of the microscope – taking the place of

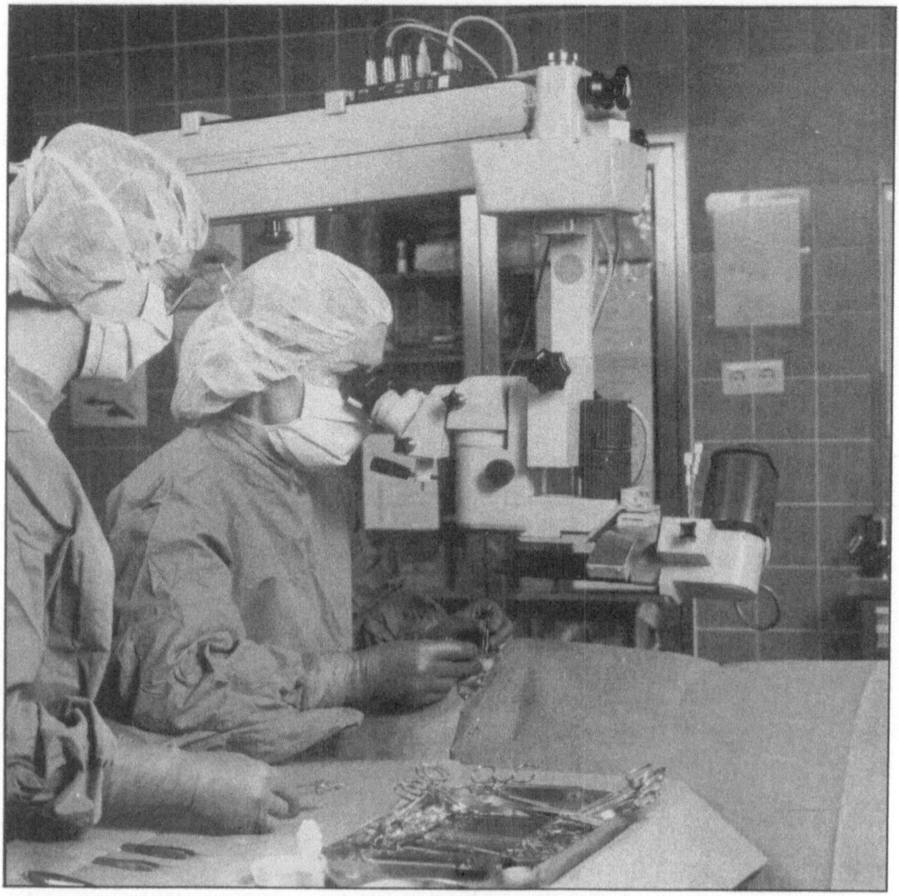

Fig. 4.5.   *Voice input controls focus, zoom and movements of this laser microscopic equipment in use in micro-surgery*

up to 12 foot controls and improving the ergonomics and the safety of the operation.

In the USA a number of surgical applications are under development, including spoken requests for data from instrumentation, endoscopy reporting and the voice control of robotic devices.

EDUCATIONAL AIDS

To date, applications of speech technology in education are somewhat restricted and not very imaginative, perhaps partly because of the relatively high cost of hardware and software. In the USA there are applications in the administrative area where needs are similar to those in other sectors. Universities have set up voice messaging and voice mail services. They have set up telephone information services using systems such as DECtalk, including facilities giving up to date information on courses and available vacancies. They have even gone farther and provided facilities for students to register on-line using the DTMF keypad.

The keypad (alternatively speaker-independent recognition could be used) provides the means for selecting what is needed from a large volume of course information, by asking the caller to press different keys to get information on different subjects.

A telephone voice response system using DECtalk to augment a medical correspondence course was introduced in 1985 by an American organisation in association with a psychopharmacology education module of the American Psychiatric Association. Students would fill out the traditional answer sheets, but instead of returning them by mail they would call this service and enter coded answers via the telephone keypad. They would get an immediate response indicating whether or not their answer was correct and offering a detailed explanation of the correct answer. The particular subject was selected because it is one which is developing rapidly and the voice system gives the opportunity for frequent updating of the material.

Education modules for day schools in the USA are produced by the electronic publishing division of Jostens Learning systems. The company had developed microcomputer diskette packages using text and graphics for use on Apple II computers, which are widely used in US schools. Then, following the theory that learning was better if more senses were involved, they developed a high-quality speech synthesiser board to go into the Apple computer, together with a separate amplifier and speaker system, and introduced teaching modules which incorporated good quality synthesised sound on the same 140K 5¼-inch diskettes. The speech board had its own power supply, co-processor and 64K of random access memory, so that it did not make any demands on the main microcomputer. All of the speech software was downloaded to the 64K RAM at the start of use, the 64K being adequate to hold 5½ minutes of continuous speech at a sampling rate of 1200bits/s, which is quite adequate for a 15–20-minute teaching module. Modules developed included reading skills, mathematics, science and thinking skills.

Today, there are many teaching packages for microcomputers that incorporate synthesised speech, and the number on the market is increasing rapidly. First Byte is a US company which has developed a range of educational products for use on computers with built-in or add-in speech chips. Typical is 'Speller Bee', a program for children of 5 – 10 years old, aimed at improving their spelling skills. Children can create 32 of their own spelling lists, with each list containing 10 words of up to 15 letters. The program contains 150 pre-stored spelling words each at a different level of difficulty. 'Mathtalk' is designed to improve skills in addition, subtraction, multiplication and division. It allows children to enter their own problems and obtain individualised graphic and spoken help specific to each problem.

A consortium including the National Geographic Society, Apple and Lucas-film has introduced a package covering American geography with still and motion pictures as well as sound, all stored on a laser disc. The compact disk is a very convenient medium for storage of combined speech, text and graphics software – disks can hold up to 30 hours of speech.

LANGUAGE TRANSLATION

Simultaneous translation between languages is a demanding task for skilled linguists, and there is a large volume of research aimed at generating computerised translations. The biggest problem is the linguistic one of actually making the translation, but if input and output are to be spoken there is also a speech technology element which, for voice input, introduces all the problems of recognising spoken messages that have already been discussed.

NEC in Japan demonstrated in 1985 a system which translates a telephone call between English and Japanese and between Russian and Japanese. The system is speaker-dependent and will only work within a tightly circumscribed vocabulary.

British Telecom is also developing a language translation system which will operate between a number of languages. It is described in more detail in Chapter 8.

AIDS TO THE HANDICAPPED

There is considerable potential in both synthesis and recognition of speech for assistance to people with a great variety of handicaps and disabilities. For those who lack the power of speech, synthesis offers the possibility of more direct and immediate communication than written or typed messages, and gives such people the ability to use the telephone. Speech synthesis can also give considerable assistance to blind people by the use of spoken instead of written messages and instructions. Speech recognition brings the opportunity for voice control of all sorts of equipment – from switching on the radio or drawing the curtains to operating a typewriter – to people with major physical disabilities. It can also be used therapeutically with voice response for people with speaking difficulties. The main practical obstacle in the way of some of these applications is the cost of the equipment.

For the blind or partially sighted, in a world where so much depends on being able to read written instructions, the ability to convert written text to speech is a great advantage. There are a number of text to speech systems which work from computer-generated ASCII text, but most written matter is printed or hand-written. Optical character-recognition systems are on the market which can read a wide variety of printed or typewritten typefaces, though for most individuals their price at present would be prohibitive. In principle, though, they give the ability to listen to newspapers and books read aloud.

There is an increasing volume of news and other material available in computer databases and accessible through electronic mail and teletext services, and there are some intrepid blind people who carry on businesses by using such facilities. DECtalk have developed a prototype Talking Word Processor which gives full word processing facilities for blind users. The typist uses the keyboard in the ordinary way, but spoken help messages are provided to give guidance on editing the text, and the user can move the cursor around the text to hear what has already been typed. For editing and proof-reading there is a facility to check capitalisation, spacing and punctuation, and different synthesised voices are used for different functions so that the user knows which mode is in use. The system is under evaluation by a number of visually impaired people.

For speech-impaired and non-vocal people, text to speech offers a partial solution, a major hindrance being that the text must either be typed on the spot for an impromptu conversation or must be selected from a menu of phrases and sentences, all of which makes communication slow and somewhat artificial. There have been three main approaches to the speeding-up of text input – the use of a special keyboard with ideograms to represent messages; the adoption of a system of abbreviations on an ordinary typewriter keyboard; and machine shorthand such as Stenotyping. Any of these systems would clearly need to be learned to a high level of skill before they could be used fluently.

A great deal of ingenuity has gone into the design of systems which enable physically handicapped people to carry out many tasks under voice control. At its most basic a speech recogniser uses a spoken word as a switch, so that anything which is susceptible to electrical or electronic control can, at least in principle, be voice controlled. A voice-controlled wheelchair has been demonstrated by Kempf of Strasbourg, and the same system has been applied to subsidiary controls on a motor car such as operating the window and switching on the radio, but it would be extremely hazardous to attempt to steer a car by voice control.

There are also systems for controlling equipment around the house and robotic aids by voice input, and the computer itself is of course amenable to voice control. The 'speech typewriter' is not yet a practical possibility for the office environment, but voice input is well worth considering as an alternative to such devices as puff tubes and optical head pointing for controlling word processing and other software.

CONSUMER PRODUCTS

Spoken messages are now being used in many applications as an alternative to visual displays to give brief instructions or warnings. A speech chip avoids the need for additional warning displays where there may already by extensive instrumentation. There is no problem of difficult lighting conditions which can hamper visibility of a visual display. And if a number of fairly lengthy messages are to be presented, the alternative of a cathode-ray tube monitor or a dot-matrix display will probably be more expensive than a simple speech synthesis device.

A very well-publicised application some years ago was the use of a Hitachi synthesiser to deliver messages such as a seat-belt warning in the Maestro car. In motor vehicles, as in aircraft and similar situations, there is an added reason for using spoken messages in that they do not distract the driver's attention from the road. They are also less easily overlooked than warning lights. A woman's voice was used in the English, French and German versions of the Maestro. Male Italian drivers, however, strongly resented being told what to do by a woman, and in Italy a man's voice had to be used.

Several other automobile manufacturers experimented with speech synthesis, but today Renault is the only company to offer a speech option, using a TI speech chip. There are real and practical opportunities for in-car speech synthesis to give warning when eyes are busy – and there is evidence that drivers often fail to see or to heed visual warnings – but the market has probably been made more difficult as a result of early implementations which savoured too much of gimmickry.

Brother Industries, the Japanese sewing machine manufacturer, when it introduced the advanced Compal Galaxie microprocessor-controlled sewing machine in 1982, added to its many facilities a speech output system. If the user makes a mistake in using the machine, it speaks one of 10 sentences such as 'Please change position of red knob to right to raise feed dog', or 'Reverse stitch cannot be obtained using this pattern'. According to Brother Industries the speech output electronics added 3.8% to the price of the sewing machine. A separate speech ROM was produced for each of seven languages.

There are now many applications of simple speech chips to sophisticated toys for children, and also in low-cost spelling and foreign language aids.

SPEECH TO TEXT

Some companies in the USA, notably IBM, Kurzweil and Speech Systems Incorporated have been working for some years on the idea of a system which could act as a sort of robot shorthand typist – the busy executive would dictate a letter and the voice typewriter would type it. IBM has throughout remained cautious about its achievements, but some others have made promises and prophesies which, in 1988, remain unfulfilled. In the UK there is an Alvey project along similar lines which has also had its objectives somewhat scaled down.

Nevertheless, some progress has been made, and Kurzweil is now marketing two dictation systems aimed at very specialised medical markets. One is for producing radiology reports, and allows doctors to dictate, edit and print reports very quickly. The other is for use in hospital casualty departments and allows doctors to dictate and print patient records. Fig. 4.6 shows a Kurzweil Voice EM system being used for dictation by Dr. A. L. Chambers, director of emergency medical services, Nash General Hospital, Rocky Mount, California. Reports are dictated and edited directly on the computer screen, and can be printed out immediately, removing delays in transcription and problems with illegible handwriting.

Both systems rely on the fact that the reports are highly standardised, and that the vocabulary, though large, is highly specialised with many long words which are fairly readily distinguished from each other. The basic systems have a vocabulary of 1000 isolated words, and there is a version of the radiology system which can have its vocabulary extended up to 10 000 words.

Fig. 4.6. Dr A. L. Chambers dictates emergency medical reports into Kurzweil Voice EM system for immediate printout (Courtesy of MLS/A: Ben Casey)

Brigham and Women's Hospital in Boston, Massachusetts is running a pilot project using Kurzweil recognition equipment to supplement the existing central dictation system which doctors are using to dictate their reports through the hospital telephone network. The dictated reports are transcribed off-site and delivered by courier to the medical record department, and it is becoming increasingly difficult for the system to keep pace with the demand for fast turnaround and accessibility of reports.

The project began with the help of three surgeons, each of whom selected a procedure, and their previously dictated operation reports were used to obtain a count of all the words which they had used. The procedures were breast biopsy, herniorrhaphy and cholecystectomy, and there were found to be 661 unique words in the three reports – an average of 220 words per procedure. Each surgeon trained the computer in the words appropriate to his or her subject – which required about 90 minutes.

Because reports contain a number of regularly used sentences and phrases, it has been possible to employ single key words to initiate the entry of quite lengthy pieces of text, as in Fig. 4.7. The method has achieved its objectives of eliminating transcription, standardising operation reporting and reducing the time needed to produce reports, but it has yet to be discovered whether the system can be used effectively throughout the hospital and whether it will be acceptable to other surgical staff.

	Key phrases – Cholecystectomy
Say:	*Narrative display:*
Severe episodes	*The patient has had severe episodes of acute cholecystitis lasting (digit) hours*
Supine	*The patient was placed in a supine position and the abdomen was prepared and draped*
Subcostal	*A subcostal incision was made*
Single stone	*An ultrasound was done and showed a single stone*
Ties	*Haemostatis was obtained with ties*

Fig. 4.7. The use of key words and phrases to represent standard sentences in dictated reports

5 HUMAN FACTORS

'READING maketh a full man; conference a ready man and writing an exact man.' So wrote Francis Bacon some 400 years ago, pin-pointing the fact that talking is fine for fast communication, but it leaves something to be desired where accuracy is important. If this is true of communication between people it suggests that extra care is needed when the listener is a machine.

Take the simple case of inputting commands to a computer. Most commonly this is done with a special keypad or a conventional QWERTY keyboard, perhaps with the addition of special function keys. For some purposes it is also possible to use icons on a screen together with a mouse or a touch screen, and where graphical information is to be conveyed, as in computer-aided draughting, a tablet or light pen may be used.

A spoken word or phrase can be identified by a speech recogniser and used in the same way as a function key or a typed command. Spoken numbers from 0 to 9 can perform the same function as pressing the keys 0 to 9 on the keyboard, and with greater difficulty the whole upper and lower case alphabet might be represented by speech commands. If there is a simple choice between speech input and a keypad or keyboard the vital questions are:

• How fast is it?
• How reliable is it?
• What does it cost?

Unless the spoken input takes the place of a lengthy typed message there is probably little significant difference in the time required for speech and for keyboard input. A bigger difference will show up if verification is required. A typed entry can be checked as it is entered, and an error will probably be seen and corrected immediately. A spoken instruction can be checked in various ways. A message can be written to the monitor screen, and if wrong a spoken correction can be applied. If there is no screen, a synthesised message can be played back – 'Did you say . . . ?' The greater the accuracy required, the slower is speech input likely to be compared with keyboard input. Even where the speed and accuracy are comparable, the greater cost of speech equipment compared with keyboard data entry is likely to make it less attractive unless other factors are involved.

One exception is where a large number of special functions are required. A marginal case is word processing, where there is a choice between either an extended keyboard which has to be searched for function keys or the use of 'control characters' using multiple keys which must be memorised. Spoken input has been used to supplement ordinary typing with commands for cursor movements, block movements and so on.

More clearly advantageous is the use of speech input at computer-aided design – CAD – terminals where very large numbers of commands are used and the operator prefers not to look away from the screen to the keyboard or tablet. This is a simple example of the 'eyes or hands busy' situation where speech technology offers a viable alternative. An even more worthwhile case is in the production of maps and charts.

If the operator does not have access to a keyboard, or must walk some distance from the workplace to a keyboard, or where the alternative is to carry around a heavy hand-held terminal, the case for speech input becomes much stronger, and many of the practical examples in this book relate to such situations. The same is true where the only means of communication is a telephone or a radio telephone. The telephone keypad can be used as an elementary data input device, particularly where DTMF dialling is in use, but for more complicated communications there is a strong incentive to look at speech recognition.

The decision on economic or operational grounds that speech recognition is a worthwhile technology is only the beginning of the human problems associated with speech input/output. Let's look at some of the types of task where speech input might be considered.

- *Eyes and hands busy.* If the user is already fully occupied – perhaps, driving a car – what will be the effect of giving him or her a task that uses another means of communication such as speech? This is a subject which has been researched in some detail because of military interest in the best way of dealing with very complex tasks. It seems what when a person has to carry out two tasks simultaneously there is some loss of efficiency in both. However, when one of the tasks involves spoken input the overall performance is better than if both tasks require manual input. This is probably because brief spoken messages cause divided attention for a shorter period than is caused by control of manual tasks.

  Most industrial and office tasks do not require the same order of concentration as military aviation, and a momentary loss of attention does not usually have such serious consequences. Two useful conclusions can be drawn, though, that there is some loss of performance when simultaneous tasks require close concentration, but the loss is less when one of the tasks uses spoken communication.
- *Type of user.* It is important to take account of the type of person who will be using the speech system, and whether the system will be in use frequently or only occasionally. The user's motivation is a key factor, and adequate time must be allowed for training.
- *Environment.* The environment in which the user has to work can make an important difference to the suitability of speech input.

People who have to use protective clothing and gloves find keyboards difficult to use. If they have to walk about over a large area or if they have to clamber over large fabrications, for example, and need both hands free, it is not easy for them to use a hand-held terminal. A very noisy environment may hinder the use of speech input, though this is a problem which now seems to have been overcome, as some of our examples indicate.

• *Technical limitations.* Current limitations of speech recognition systems, coupled with over-high expectations about what they can do, can lead to frustration and rejection of the whole technology, so great care is needed in the design and implementation of speech systems. Difficulties come from two main causes – the need for precise diction by the user, and the annoyance resulting from repeated misrecognitions.

The need for clear and consistent articulation, with each word spoken separately, is a distraction to the speaker until it becomes habitual. Even with a connected-word recognition system each word has to be fully articulated, which is not a natural way of speaking. The technique is one which must be taught, and extended practice may be needed before it is fully mastered.

Mistakes in recognition can be frustrating if the same mistake is repeated, and this can lead to further trouble. After saying the same word twice and having it misrecognised, the speaker is liable to become slightly irritated and change the pitch or volume of the voice, making recognition more difficult. So a recogniser which fails repeatedly on one or two words can be far more troublesome than one which frequently fails the first time a word is spoken but usually picks it up the second time.

Difficulties are also increased if there is no visual feedback indicating that the message has been correctly received. Even between two human speakers a task such as dictating a name and address over the telephone can be a lengthy and irritating task. If confirmation is necessary using synthesised speech, time has to be allowed for the message to be received, checked, repeated back, confirmed and processed.

Good speech system software should provide a number of standard facilities to help accurate data entry, including:

• User prompting.
• Syntax evaluation.

- Feedback.
- Editing strategies.
- Output buffer maintenance.
- Conversion between spoken language and coded representations.
- Help messages.
- Easy training and retraining of utterances.

THE IMPORTANCE OF DIALOGUE DESIGN

The foregoing paragraphs point to the need for careful design of the structure of dialogue between humans and computers so that it is simple and accurate, requiring the minimum of corrections and feedbacks, and so avoiding annoyance and frustration.

A useful measure of the efficiency of a speech interface is the 'transaction time'. This is defined as the time taken from the user's receipt of a stimulus to carry out a task to the point at which the user is satisfied that the task is complete. The starting point may be a synthetic speech prompt, say to look for a particular characteristic or defect in an inspection application, or it may be the inspector's own decision to initiate a task. The transaction time is complete when the relevant information has been entered into the computer and the user is assured that it has been entered correctly. An inefficient transaction may be delayed by an ambiguous prompt or one which does not follow naturally from the previous one so that, for example, an inspector has to move from one activity to another and then back again. Too large a vocabulary may cause the speaker to forget the correct response or to fumble for the right word, and may also contribute to too high a level of misrecognitions, causing further delay in correcting the data. A well-designed dialogue can play an important part in minimising transaction time. Fig. 5.1 indicates the structure of a simple dialogue design.

Specialists distinguish between two kinds of 'dialogue acts' – *goal-oriented acts* in which the computer asks or is told what the user wants it to do, and *dialogue control acts* which are more concerned with preventing and correcting errors in the dialogue.

The important thing in a goal-oriented act is to ensure that it is clear and error-free. A new or infrequent user of a system may have to be given more detail than somebody who is using it regularly, and there may have to be provision for bypassing the help facilities

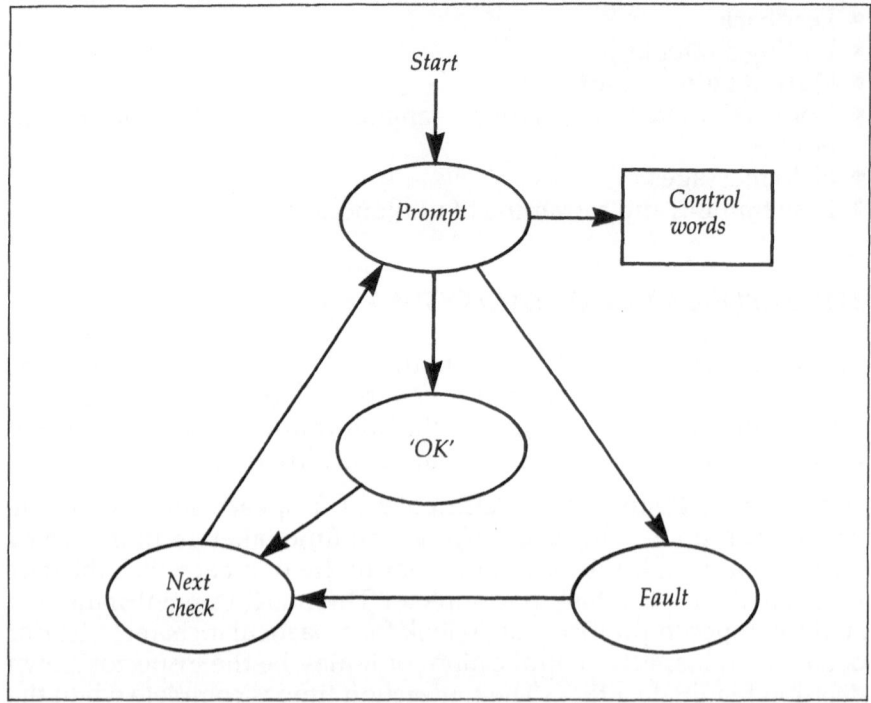

Fig. 5.1. Simple dialogue design diagram

if people with different levels of experience are to use the same system. As with ordinary computer software there are several ways of carrying on a transaction, including command languages, menus and question and answer sequences. Experienced users generally prefer to control the transaction with direct commands, because they do not need to wait for menus to be presented to them, or questions to be put.

An experienced user can employ a command language to avoid going through unnecessary parts of a routine. An inspector, for example, can simply name the defects that have been found, without having to go through the formality of reporting every other characteristic as satisfactory.

If there are only a few words which a recogniser can expect to hear at any one time, its accuracy in distinguishing them will be much greater than if it must select a word from a large vocabulary. One way of minimising the vocabulary is to design the system as far as possible with a fixed sequence of questions, for each of which there are only a limited number of answers. Alternatively, dialogue should be designed in a tree structure with a number of branches. In this way, a large vocabulary can be divided up so that only a small part of it is usable at any one time.

A visual display makes it possible to adopt a quite complicated menu structure, but if the user can only receive spoken prompts it is dangerous to place too much reliance on the user's memory. If menus are used they should be kept short. It is probably better to give the user the initiative in speaking commands, supported by help messages when requested.

Some sort of feedback is usually desirable to verify voice input. If it can be displayed on a screen, this is the quickest and most reliable method. Voice feedback is slower and more prone to misunderstanding. Two levels of feedback are distinguished – *primary* and *secondary*. *Primary feedback* takes place if the system responds directly to the user's command and then, if the effect is not obvious as with a telephone database update, it tells the user what it has done. It is used where errors are not critical or where the time saved outweighs the cost of possible errors. It tends to be used in applications such as package sorting, where the small percentage of errors does not justify slowing down the whole process by introducing a verification phase, especially as there will be later opportunities to detect and correct the error.

Even with primary feedback, though, it is worthwhile to have a 'word confirmation dialogue' where the recogniser finds two words that fit the spoken utterance almost equally well. Since this only happens occasionally if has no significant effect on the overall time for data entry, but has been found to improve overall accuracy by as much as 5%.

In *secondary feedback*, the command is repeated back to the user for verification before any action is taken, and this method is preferred where accuracy is particularly important. Examples are where the information is of permanent importance as in quality control, or where an error can lead to dangerous consequences as in control of an industrial robot.

Where feedback can be visual, for example on a large screen within view of the operator, it is preferable to spoken feedback as long as it does not cause too much distraction from the work in hand.

Error correction needs some careful planning if several pieces of information are to be entered together. For example, the inspection of a motor car body may require, for each defect found, a description of the general area where the defect is located (boot lid), a more exact position (top left), the nature of the defect (scratch), and an assessment of its criticality (important). It may be desirable to play back the complete message at the end of the sequence, unless there is an uncertainty in a recognition – 'Did you say scratch?'.

The type of editing for errors which is adopted seems from experience to depend on whether the playback is to a display screen or is spoken. With a spoken playback the preference is usually to cancel the whole message and start again. With a visual display it may be preferable to have spoken editing commands which allow the user to step through the message and single out the misrecognised utterance for correction.

CHOOSING THE VOCABULARY

There are two main considerations in deciding on the vocabulary to use in a recognition system, and they can conflict with each other.

First, the vocabulary must as far as possible be a natural one for the user. For an inspector it should use the words which would normally be employed in describing a defect. The more appropriate it is in the mind of the user (not necessarily to the software writer), the more likely it is to be remembered. The same applies to the command words like 'repeat', 'erase', 'next', which have to be available in every vocabulary.

The other consideration is that the danger of confusion must be avoided as far as possible. The word 'next' mentioned above is a case in point. Some users have found that it is confused by the recogniser with 'six', and it is necessary to change to an alternative such as 'advance'. Alphanumeric characters cause particular problems. 'Five' and 'nine' are notoriously similar. 'B' and 'P', 'D' and 'T' are easily confused, as are many other letters of the alphabet and it is virtually essential to adopt a phonetic alphabet. In fact where lengthy alphanumeric strings have to be entered there is a strong case for adopting bar coding for data entry of such things as part numbers. Otherwise it may be faster and more reliable to enter alphanumeric characters at a keyboard.

One useful rule in avoiding confusion is to keep the number of words in a single vocabulary as small as is consistent with the needs of the immediate task, and to use one of the voice commands to switch to another vocabulary. In this way many of the clashing words can be separated, because the recogniser will only look for matches to the utterances in its current vocabulary.

TRAINING PROCEDURES

Speaker-dependent recognisers have to be trained by each user to recognise every word in the vocabulary spoken by the user. Some recognisers, under favourable conditions, require each word to be spoken only once. Others use a method requiring multiple patterns for each word. Depending on the recogniser, the speaker and the words taught, anything from one to eight utterances of each word or phrase may be necessary.

One of the biggest difficulties with users coming to speech recognition for the first time is in persuading them to speak naturally. People tend to be self-conscious in speaking to a machine, and nervousness affects the sound of the voice. They may shout at it or speak very quietly. The most common variation is to articulate each syllable very clearly as though to a foreigner. Small wonder that the recogniser fails to respond when the speaker relapses into his or her normal voice after a few minutes.

This process of settling down to using a recognition system takes time. If the users work in a shop floor environment alongside other people it may be advisable to give them their introduction to using the system away from the shop in something more like a laboratory atmosphere, where they will not feel so self-conscious. When the system is first put into operation in the shop there will probably be further unwelcome interest and amused comment for a few days. Before long, though, the new technology becomes part of the normal life of the organisation and companies usually find that proposals for second and third applications are enthusiastically welcomed.

Initial training of the recogniser to an acceptable level may take an hour or so, depending on the size of the vocabulary, but the number of misrecognitions will be much higher than the claims in the manufacturer's literature. At this stage there will have to be fairly frequent retrainings of the recogniser as the speaker's voice settles down to normality. The learning curve will vary between users, but it may well take a month of regular use before very high levels of accuracy are reached.

From time to time after that there will need to be retraining of individual words or groups of words. People's voices change over time, and are affected by tiredness, colds and other factors. But some regular users of speech recognition claim very high levels of first-time recognition – such as one mistake in 4500 utterances.

QUALITY OF VOICE RESPONSE

Speech is a highly personal activity. We rely on a voice to tell us a great deal besides what is actually spoken. Without seeing the person we can draw conclusions – not always correct – about sex, age, nationality or regional origin, whether the person is friendly, unfriendly, angry, worried, inebriated and so on. Therefore, we cannot avoid making similar unconscious judgements about a synthesised voice, even though we may recognise it as synthetic. This leads to two interesting and important questions about choosing a synthesised voice.

- *Should the voice sound natural?* In an interactive situation where the hearer could mistake the voice for a human one it is wise to maintain some degree of artificiality about the voice while ensuring that it is easily understood. Otherwise the quality can be as good as the technology and the storage capacity of the system will allow.
- *What sort of voice is preferable?* The most obvious difference is between a male and a female voice, and the choice is very much a subjective matter. In a telephone voice response application the user may expect to hear a female rather than a male voice. In a frequently used interactive situation such as an inspection task, any particularly idiosyncratic features should be avoided, but a regional accent may be appropriate. In most practical applications the necessary compression of the synthesised speech signals is sufficient to remove any characteristics that might identify an individual speaker. Text to speech applications are at present limited by the technology to a rather artificial presentation, though quality is improving. Where the text is specially prepared for reproduction there are various ways in which the quality can

be improved by accenting, pauses and other devices, and where a word is not included in the system's pronouncing dictionary it may be possible to 'phoneticise' the spelling.

6 HOW TO GET STARTED

THERE ARE many companies offering speech recognition and synthesis products and services, and the range of applications they cover is very wide. In certain defined application areas like voice mail and telephone enquiry systems there are standard products and existing applications which make installing a new system, a relatively straightforward task.

Industrial applications involving speech recognition, however, are much more varied and each one introduces new features requiring special attention. It is here, particularly, that important decisions need to be taken about the best way to approach a prospective speech application.

TAKING ADVICE VERSUS WORKING INDEPENDENTLY

It is possible to buy speech recognition boards and equipment for use with, for example, a microcomputer. The cost is not great, and some companies have bought such equipment to be tried out by the manufacturing systems people. Simple for familiarisation with the technology, this is quite a useful exercise. However, there is a very large gap between getting a recogniser to respond to a small spoken vocabulary and implementing a reliable speech system as an integral part of a quality or logistics programme.

For perhaps £1500 – an amount which counts as petty cash in many firms' budgets – you can buy a speech kit for installing on a personal computer. But partly because its cost is low, the equipment does not attract the close attention of senior management and does not give the sense of importance in achieving results that would be demanded of a capital project. All too often, the engineers will experiment with the equipment for a time, discover that recognition is not 100% reliable, and eventually drop the investigation, realising that there are many difficulties and pitfalls in working towards a practical system. This negative experience has been known to deter companies from looking afresh at the potential for speech recognition, and to delay worthwhile implementations for a matter of years.

For this reason it is nearly always best to approach a potential speech application in collaboration with a company which already has experience in applying speech technology – either with its own proprietary system or as an independent systems integrator. Anybody setting out for the first time to implement speech recognition and synthesis in an industrial environment will make a lot of blunders and will have to learn many lessons the hard way. It will almost certainly prove more economical to go to a company which has already learned lessons from its mistakes, and to profit from that experience, than to fall down all the same holes oneself in tackling the project alone. You will have to pay for the benefit of that experience, but the final cost will almost certainly be less than that of carrying out your own development, and there is much greater likelihood of a successful implementation at the end. After a first speech project has proved its worth and you have gained experience with it, you may well decide that you have acquired the skill and confidence to tackle a second one with your own company resources.

If an investigation of data capture methods in your company reveals one or more areas where speech technology appears to offer advantages, you could approach some speech systems companies and ask if they would be willing to mount demonstrations to give an indication of how such an application might work. Alternatively, you could ask an independent consultancy or speech company to come in and make its own assessment of potential applications and to mount a demonstration. If the proposed application is similar to one already implemented in another company, it may involve minimal writing of new software, in which case you could expect the work to incur only a small fee or even to be free of charge. You can also ask to see any existing implementations with similar features to your proposed system.

Because the whole field of speech recognition applications is new, you and your prospective suppliers may have difficulty in finding existing implementations which correspond closely with what you have in mind, in which case rather more work may be involved in preparing and mounting a convincing demonstration. This is especially true if the speech system is to be simply one data capture element in a much larger system such as quality control, production scheduling or order processing and you may have to commission a pilot study leading to a demonstration. In that way you should be able to clear up any uncertainties about its feasibility.

Depending on your supplier, money spent in mounting a realistic demonstration may be recoverable in the cost of a subsequent full implementation. At current UK prices, a feasibility study leading to a realistic demonstration might cost something like £10 000, which for many companies would come within the scope of revenue spending. A full implementation, including the hardware and software, interfacing to a database, graphic displays and so on, might cost in the region of £20 000 – £25 000, with a further £10 000 for consultancy and support.

The basic speech hardware and software represent a relatively minor part of the total cost – perhaps between 25% and 30%. The major part of the cost goes into ensuring that the speech system supports the user. In an industrial application, particularly, it involves:

- Careful dialogue design.
- Thorough familiarisation and training of the people who will be using the equipment.

- Tuning of the system to ensure that the information entered is appropriate, adequate, accurate and in a form suitable for interpretation, for example by the quality control department.

Subsequent maintenance and support of the speech system does not make great demands on a company's resources. The equipment is similar to any other electronic and computer equipment in the company. The speech vocabulary, both input and response, will need to be updated as products and processes change, and some knowledge and experience in dialogue design needs to be maintained within the company unless all such work is to be passed to an outside contractor. In any company with a major commitment to computer usage there will be a systems department, and this would be the natural home for any expertise in the design and development of speech systems.

Systems people will also be involved if a company decides to develop its own speech system for data capture, but the departments directly affected by any change will need to be closely associated with the project.

WHERE TO LOOK FOR APPLICATIONS

Whether you decide to use the services of an outside organisation or to develop your own application your first step will be to look for the areas where speech technology could help to solve some of your current problems.

Industrial speech recognition applications are all basically concerned with data capture, and speech recognition must take its place alongside other methods such as keyboard entry, bar-code reading and electronic tagging as one of the possible means of electronic data capture. Fig. 6.1 sets out some of the factors that should weigh in the decision between these alternative methods.

- *Need to have eyes free for other tasks*
- *Need to have hands free for other tasks*
- *Need for mobility, epsecially if this involves work handling or difficult access*
- *Situations requiring direct computer input*
- *Situations requiring immediate feedback or use of results from data collection*

Fig. 6.1. Factors tending to favour the use of speech for data entry

First in importance among factors pointing to speech is the need to have eyes free to concentrate on the work in hand. If close attention is particularly important, then voice input commands may be preferable to using a keyboard, keypad or machine controls which the user may have to look at from time to time in order to operate them reliably.

If eyes and hands are both occupied, there is a very strong incentive to use voice input for any further tasks which cannot be done with the hands or which would slow down the whole operation to an unacceptable degree.

Mobility can be an important factor favouring speech systems. With a lightweight microphone headset and battery-operated radio transmitter the user does not have to keep returning to a computer terminal to enter data. A hand-held terminal may be a satisfactory alternative to speech input. Information can sometimes, though not always, be entered faster, and there may be faster confirmation that the entry is correct. On the other hand, the terminal is bulky, relatively heavy and inconvenient, and must be carried about by the user.

Where protective clothing must be worn there is often a strong case for speech input because of the sheer difficulty of using a keyboard. Where appropriate, the microphone and earphones can be incorporated in the protective clothing.

A noisy environment might seem to be a contra-indication for speech input, but in fact speech recognition has been used successfully in situations so noisy that ordinary conversation is impossible. Naturally the choice of appropriate equipment for such a situation is particularly important.

ECONOMIC ADVANTAGES

So far we have concentrated attention on the operational advantages of speech input, but there are equally important economic questions to be answered. In an industrial situation a speech system must show benefits commensurate with its cost and the benefits should, as far as possible, be direct and measurable.

As our case studies show, there are applications which have resulted in direct labour savings, for example where one person can carry out a task formerly requiring two people. More commonly, the savings appear as part of a wider programme to improve the efficiency and accuracy of information flow. There are many situations today where an inspector, a progress chaser or a warehouseman makes written notes, perhaps completes a form in duplicate or triplicate or worse, and then either this form itself travels around the company providing information to the departments concerned, or the information is keyed into a computer by a clerk. Sometimes there is both physical handling of paperwork and rekeying of the information into separate computer systems.

When a company addresses this problem and looks for ways of eliminating duplication and the associated dangers of error, it has to face the question of direct data entry by the person who originates the information. Speech is just one of several options which may be considered, either individually or in combination. Most common is keyboard entry at a computer terminal, but this can be supplemented or sometimes replaced by use of a light pen or mouse. Bar coding is being used more and more widely as a means for entering fixed information such as part numbers or pallet numbers, and it can be used effectively with keyboard or speech input. Some companies have opted for a combination of keyboard, speech and bar coding to achieve maximum efficiency of data entry.

A major reorganisation of information flow in a company can lead to substantial benefits resulting from the faster and more accurate transfer of data and the elimination of duplicated effort. The results of inspection, for example, can be fed back immediately to manufacturing departments to prevent further production of defective work, and forward to ensure that rework is carried out promptly. The information can be automatically collected and analysed statistically to give an up-to-the-minute picture of quality trends, so that corrective action can be taken promptly.

Where speech input is adopted as part of the reorganisation of information flow it would be foolish to attribute the entire economic benefit to a particular method of data entry. Nevertheless, there are cases where the availability of speech technology is an important factor in the viability of the whole project.

In major integrated applications the speech element, even though a small part of the total, may involve a number of workstations and some quite complicated vocabularies. Before launching into an application of such complexity is it wise to start by developing experience on a simpler pilot project.

COMPARING SPEECH WITH OTHER METHODS

One approach to measuring the relative efficiency of speech and other methods of data entry is through the concept of 'transaction time', which has already been mentioned in Chapter 5 where human factors were discussed. This is the time from the moment a user receives the stimulus to carry out a task, to the time he decides it has been carried out satisfactorily. As we have seen, transaction time also provides a yardstick for measuring the efficiency of any particular speech implementation, and a pointer to further improvement.

The main speed advantages of speech compared with keyboard entry are found where:

- Natural language is required rather than numerical data.
- Users are not skilled as keyboard operators.
- Hands and eyes are busy on other tasks, which must be interrupted to carry out keyboard entry.

The speed of voice input is also going to be affected by:

- The user's ability to remember input commands.
- The rate of speaking.
- The accuracy of the speech recogniser and the consequent need to repeat commands and make corrections.

The first and second factors are connected with good dialogue design. Recognition accuracy of the equipment itself is related to its technical specification.

PERFORMANCE EVALUATION OF RECOGNITION SYSTEMS

One must be sure that the available technology is capable of dealing with the task. Speaker-independent recognition can only cope with perhaps 15 predefined words, and even then it will produce a high proportion of misrecognitions with some speakers and accents. The benefits must be weighed against the frustration and loss of time which may be caused – especially if the speakers are customers or potential customers.

Size of vocabulary is another consideration with speaker-dependent recognition. The larger the vocabulary, the greater the chance of misrecognitions and the greater the difficulty for the user to remember the words which are available. A large vocabulary also demands more time on the part of the user in teaching the spoken words to the system – and in reteaching words from time to time.

Some products have facilities aimed at countering some of these difficulties. It may be possible to break up a large vocabulary into a tree structure of small vocabularies with key words to initiate a change of vocabulary.

Recognisers vary in their sensitivity – the reliability with which they will identify a particular spoken sound. If the sensitivity is too high the recogniser may fail to recognise a word spoken a second time by the same person. If it is too low, the recogniser may give the same response to different words, such as 'five' and 'nine'.

Some standard procedures for evaluating the performance of recognisers would clearly be very helpful to users, particularly if the results could be expressed in a form which would allow a non-specialist to assess their relative suitability for particular purposes. Research is in progress on this subject and there is a brief discussion of it in Chapter 8.

7 INDUSTRIAL CASE STUDIES

THE GREATEST problems in implementation, as distinct from developments in speech technology itself, arise in industrial applications, because of the great diversity of problems to be solved and the need to integrate the speech function into widely differing systems. The cases that follow exemplify the kinds of applications of speech recognition that are currently going into industry. All except one are related to inspection in some form, because this is currently the most important field of application. The experience of users in installing and running these systems, however, is relevant to any situation where voice input and response are combined in a dialogue with a computer.

JAGUAR CARS

Logica has recently implemented a fully integrated system at Jaguar Cars for vehicle assembly quality assurance. One in ten cars is taken from the line for a detailed visual check of appearance and operation of the vehicle. It is an essential feature of the system that any defects found on a vehicle during this inspection lead immediately to three actions:

- The information is fed back to the assembly section where the defect originated, and operators are asked to ensure that the fault is not repeated.
- Information about the defect is fed to the final sign-off area where all vehicles coming off the line are inspected to check if they have the same defect until it has been verified that the problem has been dealt with. When the audited vehicle arrives at the final sign-off area it is put into quarantine until rectification work is completed.
- The complete inspection report is transferred automatically to the central vehicle quality database from which a variety of summary and detailed quality reports are extracted at frequent intervals by quality control personnel. Other specific investigations may be carried out and reports generated as required.

Fig. 7.a.1. Two inspectors simultaneously inspecting a Jaguar XJ-6

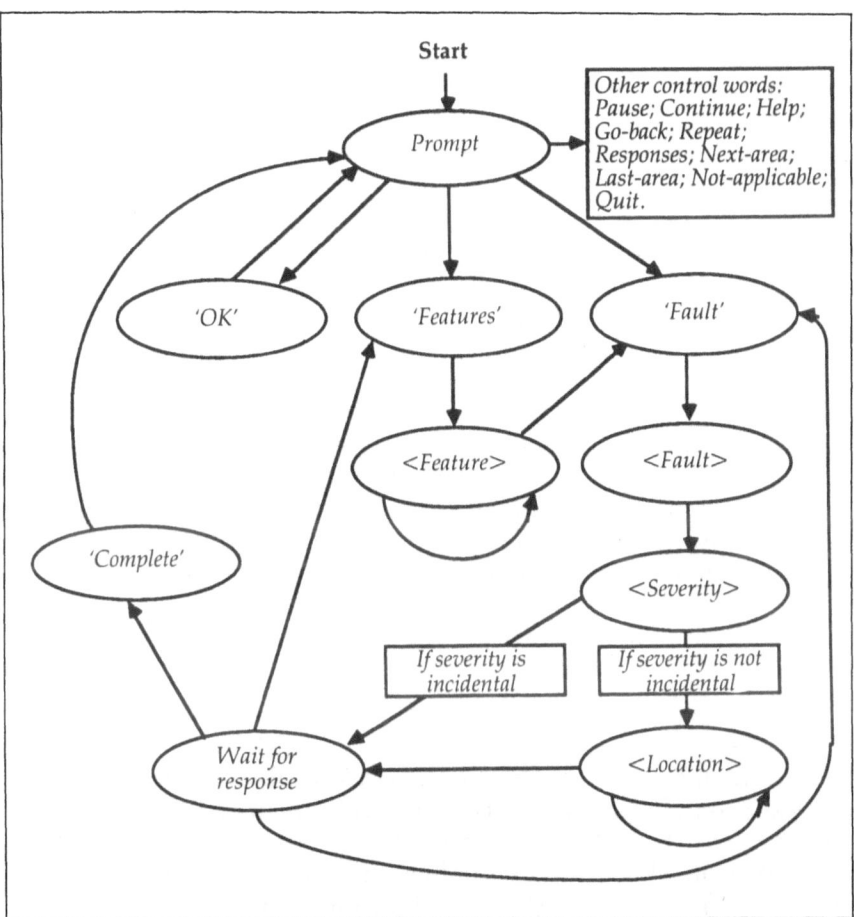

Fig. 7.a.2. Dialogue design diagram for Jaguar Cars

The quality audit is carried out by two experienced inspectors, who work simultaneously on the same vehicle. They follow a detailed checklist which differs according to the particular type of vehicle being inspected. Inspectors have to be free to get into and under the car in order to perform their scrutiny, and this is made easier and quicker by the use of an interactive speech system with a radio link to the Quality Surveillance Centre computer, which is connected via a factory-wide communications network to the central vehicle quality database computer.

The new system has made possible a level of audit inspection greater than that achieved in the past. Line inspection is carried out with the help of printed report forms which the inspectors complete and send down the line with the vehicle. Whilst this contributes to improved product quality it does not satisfy the three objectives listed above.

In 1986 Jaguar put out an invitation to tender for a quality control system to capture defect data associated with 300 checks and providing a broad range of choices of responses on a number of different models. The system was to include facilities for managing audits, collecting data from the audits, passing information back to a central facility, providing facilities for longterm data analysis and trending, and finally for managing the rectification of any faults that might be found. The method of data capture was not specified, but the company believed that it was likely to involve some sort of hand-held computer terminal, affording the inspector some free-dom of movement. Otherwise the only restriction was that the system was to use the company's local area network for communications. Logica saw it as a highly suitable opportunity for speech input and response, and put in a proposal for the complete system including a speech interface. It was the only proposal incorporating the use of speech technology, and Logica won the contract.

The speech workstations are based on IBM PCs with 2-Mbyte RAM, and 20-Mbyte Winchester disk, a 1.2-Mbyte floppy disk, and serial and parallel ports for communication. Speech processing is performed by Texas Instruments interface boards mounted in the PC expansion slots. This combines speech recognition functions with digital recording and playback facilities. Two boards are used in the same PC to allow two users simultaneous access to the workstation.

Communication between the inspectors and the workstations is via radio link, for which Logica designed and built a radio interface unit for the speech board. Each user has a light-weight microphone and headset, a voice-activated switch and a radio transceiver.

Where radio transmission is involved in the UK, the Department of Trade and Industry has to give approval. This was obtained after it had been made plain that the equipment was for a quality improvement application.

The voice-activated switch was an important part of the develop-ment. If the inspector had to use a press-to-talk (PTT) switch in order to speak it would remove much of the advantage which speech input offers through eyes- and hands-free operation. The switch is activated as soon as the inspector starts to speak. There is a more detailed description of this switch and of infra-red transmission as an alternative in the next section dealing with the Caterpillar installation.

Initially the dialogue was envisaged as a simple sequence of questions and answers in which the inspector would have a list of about 20 words like 'OK', 'repeat', 'stop' and 'continue', which would allow him to control the procedure. Then there would be another list of about 15 words to describe the faults that he found. As the system design progressed it became clear that this simple method would be quite inadequate. Audits were needed of mechanical, electrical and paintwork features, and each would require about 30 words.

The system as it now operates has changed considerably since its inception, both in the method and in the content of the dialogue. For an experienced inspector the question and answer dialogue very soon becomes frustratingly slow, so the whole operation is now controlled by the inspector, who speaks commands to the computer and so can control the speed and detail of the inspection cycle. Each part of the inspection starts with the brief prompt indicating the check which is to be carried out. If the inspector finds nothing to report he simply says 'OK' and is prompted for the next check. He will then report on any defects found, and as long as his messages are received and recognised, the only response he will get is a 'beep'. When he says 'complete' at the end of that inspection he will receive a prompt for the next check to be performed. The structure of the dialogue design is set out in Fig. 7.a.2.

The inspector can also use words like 'repeat' and 'go-back', and there is a command 'help me' which produces a more detailed description of the inspection to be carried out, on a large display screen located above the audit stand. This includes a reference to the appropriate section in the detailed inspection manual. At one stage, text-to-speech synthesis had been considered for this task, but it was considered to be too slow and not so easily understood.

The total number of prompts from the computer has now risen to 1500, though only about 850 of them are needed for any one model of car. The recognition vocabulary, which is the set of words regularly used by the professional inspectors, exceeds 300 words of various categories.

Reporting of defects is now simply structured in a standard format. If the inspector says 'feature' in response to a prompt for a new check, the computer 'beeps' and calls up the small vocabulary specific to the check. The inspector now specifies the location of the fault, using the vocabulary. If he cannot remember the appropriate word he can ask for it to be displayed on a large screen mounted overhead by saying 'responses', otherwise the computer responds with a 'beep', and he says 'fault'. Now he has a choice of types of fault, and as before he speaks the appropriate utterance (which is not necessarily a single word). When the computer responds with a 'beep' he identifies the severity of the fault with one of three words: 'alert', 'concern' or 'incidental'. His choice is important because it determines what happens as a result of his report.

Finally, if the fault is at the level of 'alert' or 'concern', he reports where on the track, in his judgement, the fault could have arisen, and this completes the entry of one fault. Now the inspector has a choice. If there is another fault to be reported against this check he can say 'feature' again, and repeat the loop. If there is nothing else he says 'complete' and receives the prompt of the next check.

Some sort of acknowledgement from the computer that it has received a message is absolutely essential, not only to provide verification but to give confidence to the inspector. Without any spoken or 'beeped' response the user quickly loses confidence that he is actually communicating and starts to think there is something wrong with the equipment or with the way he is speaking.

One of Jaguar's concerns during the implementation of this procedure was that a word might be misrecognised, and because the computer simply beeps its acknowledgement there was no feedback assuring the inspector that the message has been received correctly. Since quality was the aim of the whole exercise, the danger of data contamination was taken very seriously.

The procedure finally adopted was to use a confirmatory dialogue based on the recognition score. When a word is spoken the recogniser compares it with each utterance template in its current vocabulary and assigns a score depending on the degree to which each template matches the original word. A close match would score near zero and something very different would score near one hundred. If one word scores very much less than all the others, that is the word accepted by the recogniser, but if there are two or three templates which come fairly close together the confirmatory dialogue is introduced.

For example, 'left' and 'right' may easily sound similar, and if the inspector says 'right' and the similarity passes a pre-set threshold the computer may ask 'Did you say left'. He will respond 'reject', the computer will remove 'left' from its list. When he repeats the word 'right' the computer carries out a new comparison, omitting the word 'left'. If there is still a degree of uncertainty it could ask 'Did you say 'right'? or hypothetically 'Did you say red?' , though care is taken in building the vocabulary to avoid similar words as far as possible. If the operator had said 'confirm' the word would have been accepted and the audit would have continued.

In practice, the frequency of confirmatory dialogues is very low, and the number of misrecognitions passed through to the computer is less than 0.5 percent.

The voice response end of the dialogue, although it involves as many as 1500 prompts, is easily entered and updated. It uses digitally recorded speech which is stored on the computer. This is more natural sounding than either synthetic speech or text-to-speech.

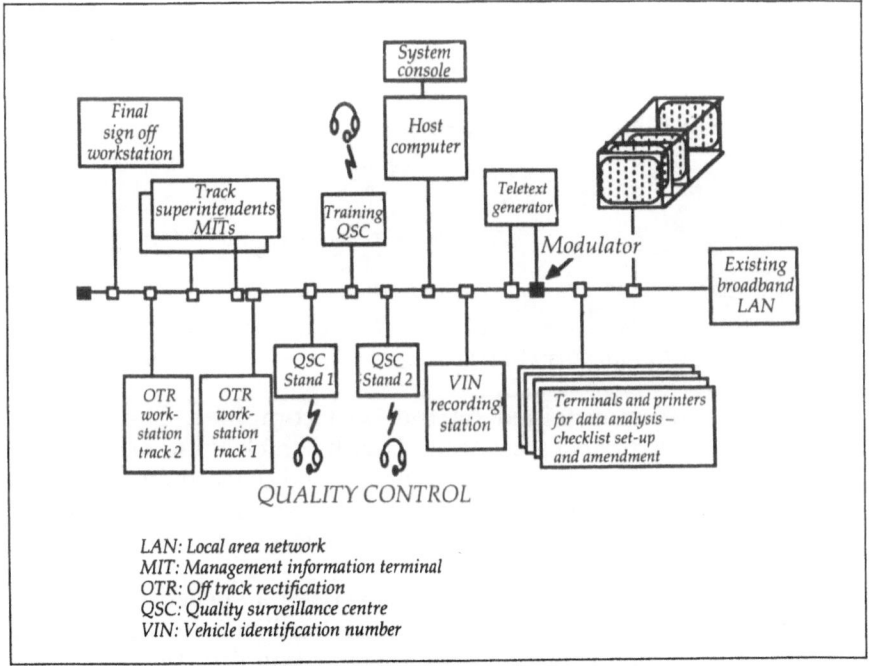

Fig. 7.a.3. Block diagram of the Jaguar quality surveillance system

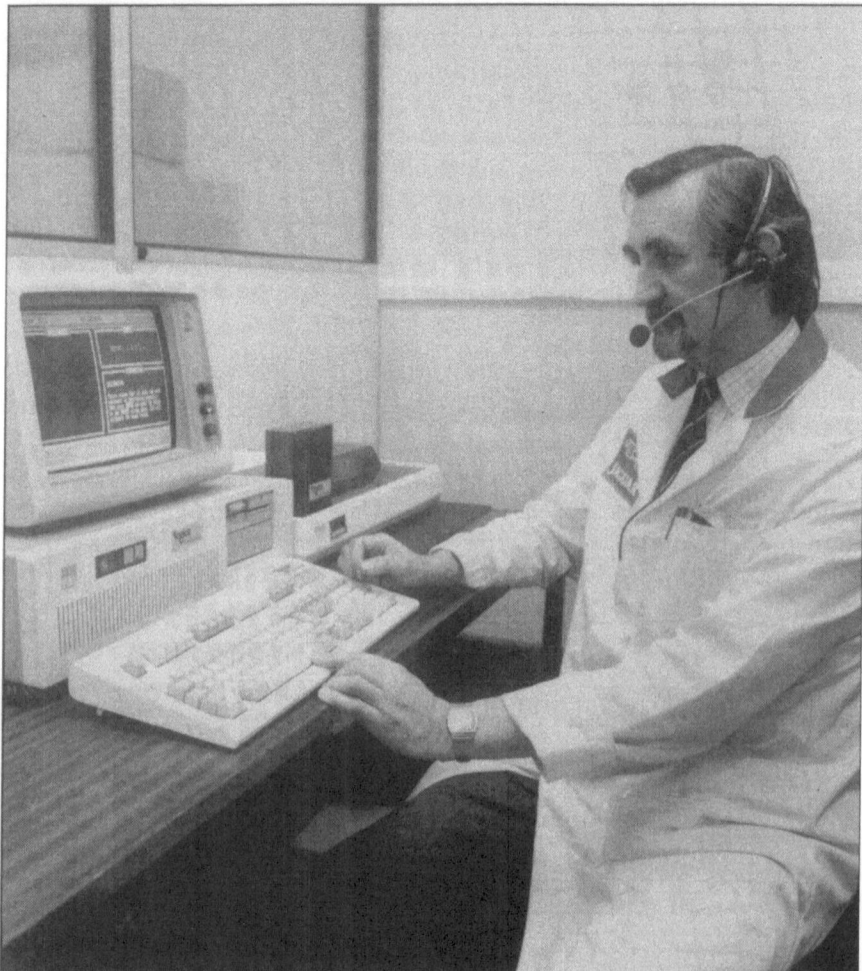

Fig. 7.a.4. Speech training workstation in use

Checklists are maintained on the host computer. The checklist structure involves checks associated with which there are features and faults. When the checklists have been validated they are loaded onto the speech workstations where the system 'learns' to recognise the operators speaking the checklist vocabulary.

Operators who had never previously used a speech recognition system took about half a day to become familiar with the training process and train up the system to recognise the 300-word vocabulary. On average, about one week's supervised use is then required to bring the operators to a position of familiarity and comfort with the speech-based system. After a satisfactory performance has been reached, further training will only require the occasional re-teaching of individual words.

The audit inspection station has been highlighted by Jaguar as a feature shown to visitors to the factory – dealers and customers – to illustrate the company's emphasis on quality, so the inspectors' reports are displayed for their benefit as well as for use in quality control.

When a car is driven on to the audit stand, the inspector uses a bar-code reader to obtain and enter into the computer the Vehicle Identification Number and other information contained in bar codes printed on the documentation with the vehicle. The model type, the build specification and the country of destination are thus identified, and the computer pulls out from its library the set of dialogues appropriate to inspecting that vehicle. If, for example, there is no sun roof there is no need to ask for checks on it. If there is left-hand drive, the appropriate prompts are needed. Jaguar has spent a considerable amount of time with Logica in developing a sophisticated matrix of checklists covering the many possible combinations of features.

At the end of the audit, a printout is produced on the stand which simply logs the faults that have been found, and this is added to the document which goes with the vehicle. Both on the audit stand and facing outwards for the benefit of plant personnel are two sets of three twenty-six-inch monitors, each displaying messages in one of three colours: red, amber and green. Above the monitors, facing outwards, is the message 'Results for audit on this track are:' and on the red screen are shown any 'alert' faults, on the amber screen the 'concern' faults and on the green screen the 'incidental' faults. The other three monitors, facing inward and placed so that they can be read easily, supply the inspectors with the 'help me' and 'responses' messages which are important to have available when there are 1500 different checks and a 300-word vocabulary available.

Information from the audit is passed over the local area network, as shown in Fig. 7.a.3, to a DEC Micro VAX II computer running the ORACLE fourth generation language relational database, where the command words are removed and the essence of

Fig. 7.a.5. Fault information is displayed on track-side monitors

the report is logged against the Vehicle Identification Number. Also on the network is another DEC computer which supports Logica's teletext system, feeding back audit information upstream to the assembly track. Large coloured monitors are mounted along the assembly track, and each monitor is tuned to a specific teletext page. The complete cycle of pages occupies about half a minute, so that updated reports of faults reach the place on the line where they probably originated within minutes of their being reported. Under the old system it would probably have been the following Monday before a defect report reached the place from where the fault emanated.

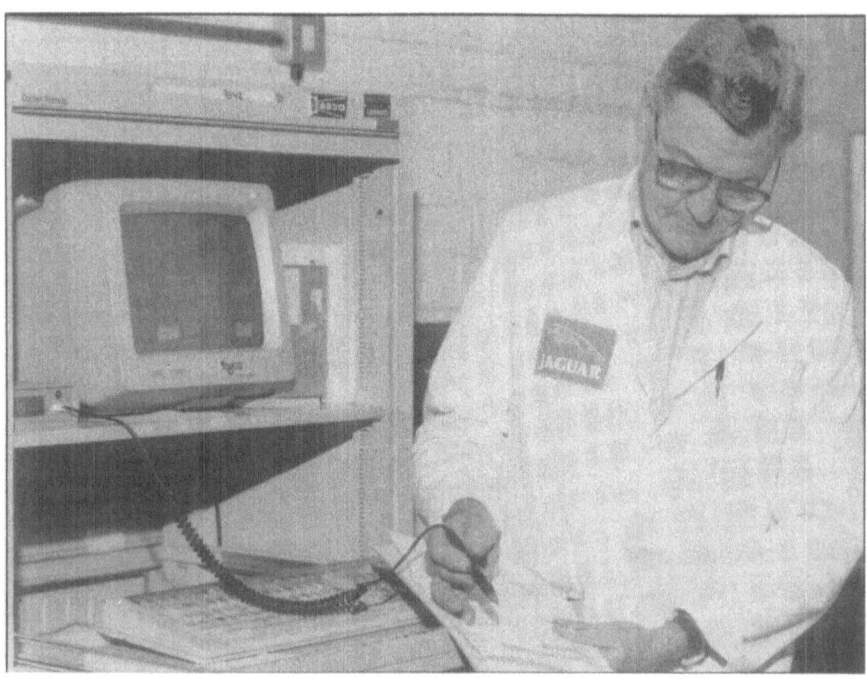

Fig. 7.a.6. Operator certifies all checks are complete

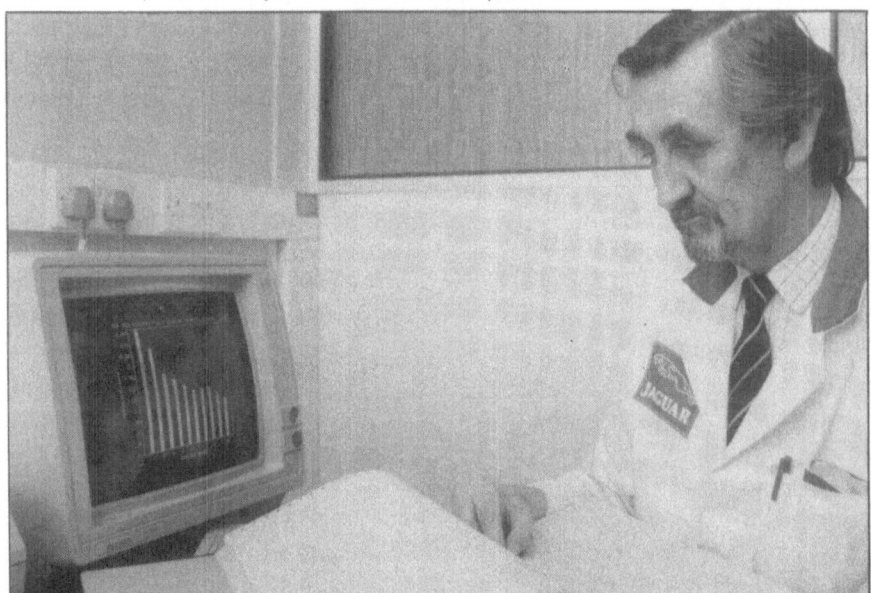

Fig. 7.a.7. Quality trends are analysed

Only one in ten vehicles goes through the audit stand, but any vehicle may have acquired defect reports from inspections on the line. The procedure at the end of the line differs a little according to whether the vehicle is to be audited or not. If it was an audit car and faults were found, it goes to the rectification area and then on to the rolling road test, where the operator can call up the computer test report and verify that the rectifications have been carried out. If it was not an audit car, the Vehicle Identification Number is read from the bar code, and a report is received from the vehicle quality database computer giving a list of check items to be examined, based on the faults found on the previous audit vehicle. Once again, these points have to be checked, along with any other reported defects, and signed off at the final sign-off station.

Finally, quality assurance personnel have access to all data acquired and processed automatically from the audit inspections, so that they can check on faults over a period of time. The ORACLE database language has an English-like query language and the database may be interrogated by non-specialists. The information can be analysed and presented in tabular and graphical form. All of this had been theoretically possible before, but it would have involved bringing together large numbers of hand-written reports, extracting the relevant information and carrying out extensive calculations. Without the reliable and regular speech-activated input to a powerful database it had been too laborious to be worth considering.

CATERPILLAR TRACTOR

On-line testing of backhoe loaders is carried out at Caterpillar's Desford (Leicestershire, UK) plant with the help of a system developed by Logica using a Texas Instruments interface board mounted in an IBM PC. The system is fully operational and working well, and Caterpillar is considering further applications of this technology (Fig. 7.b.1).

At an interim point in backhoe loader assembly the engine and hydraulics are installed. Before the tractor is completely built it is filled with oil and hydraulic fluid and 'exercised' to check the performance of the engine and the hydraulic elements. It is an extremely noisy operation. Before the speech system was installed it was carried out by two men, one of them operating the controls and the other reading the engine revolutions, hydraulic pressures and so on, and writing them down on a test report sheet which

Fig. 7.b.1. Caterpillar Desford factory with Backhoe loaders in the foreground

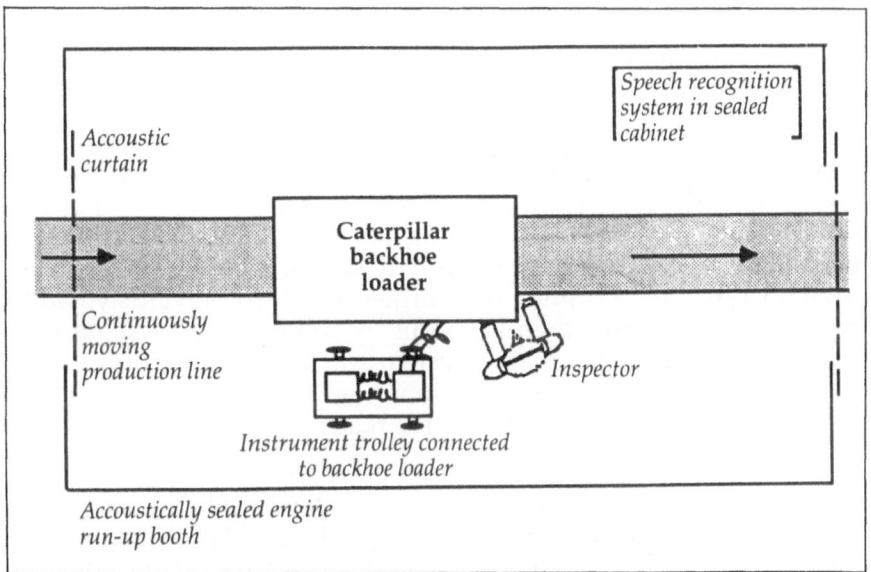

Fig. 7.b.2. Schematic layout of the on-line test area for the Backhoe loader at Caterpillar

Fig. 7.b.3. An operator using a throat microphone to report to the computer hydraulic pressures of tractor engines

remained with the tractor during the remainder of its assembly. At the end of the assembly line the sheets were collected by the quality department and used for analysis.

The test area layout is shown in Fig. 7.b.2. Twice the amount of work can now be achieved by these men, as each can speak the test results directly to a computer. The complete system cost about £25 000, so in terms of labour-saving alone it will pay for itself in two years. However the really important saving for Caterpillar came from the direct data capture of information which previously had to be written down and then keyed in by somebody else some time later. Before the speech-based system was installed, errors used to occur simply because the conditions under which the report was written sometimes made it difficult to read.

The computer still produces a printout which is a replica of the form previously completed by hand, but with all the quantities printed in. If any values are out of range the problem is logged and the printout adds a dotted line for the corrected value after the vehicle problem has been rectified. The operator then has to sign off the rectification. This part of the build cycle is similar to what was done previously, but with the advantage of speech input. The important new feature is that all the information is being logged into the PC, and at the end of the week the quality assurance department can call up a graphics software routine in the computer which plots all the readings from all the tractors and calculates mean, standard deviation and range values.

In speech terms the system is relatively simple. There are only about 20 words, and 10 of these are the numerical digits. Others are standard commands like 'go back', 'go forward', 'repeat' and 'OK'. The dialogue design diagram is shown in Fig. 7.b.4. The environment is so noisy that the headset has been built into a pair of ear defenders, and a throat microphone is used.

As with the Jaguar installation a UHF radio link is used, with a voice-activated switch. There is no electrical interference from the engine when it is running, but a substitute ignition device is used at the test station because the dashboard has not been installed, and this can cause some trouble, so users are advised to switch the engine on before using the speech equipment.

An alternative to radio is infrared transmission, which is used widely by Ford and General Motors in the USA. Because there must be a direct line of sight between transmitter and receiver there must be a number of receptors scattered around the work area. At Jaguar Cars, where inspectors are working inside the vehicles, infrared was quite impracticable. At Caterpillar it was a possible alternative and it has two important advantages – it does not have the bandwidth limitation of radio, and two channels can be open simultaneously for two-way operation. Even so the operational balance at Caterpillar was in favour of radio. Logica considers the relative feasibilities of communication technologies on their own merits in each case.

An important restriction on the radio link is that the UK Department of Trade and Industry does not favour having two channels, or even one channel, open continuously, because it becomes in effect a radio station. But even on practical grounds a continuous two-way link is not desirable. The cost and weight are greater because the user must carry a pair of radios, and the battery life is shorter.

If radio communication can be restricted to one direction with, for example, prompts to the user displayed on a large screen, an alternative would be to have a one-way continuously open channel at low power, with a radio microphone. This gives broadcast quality

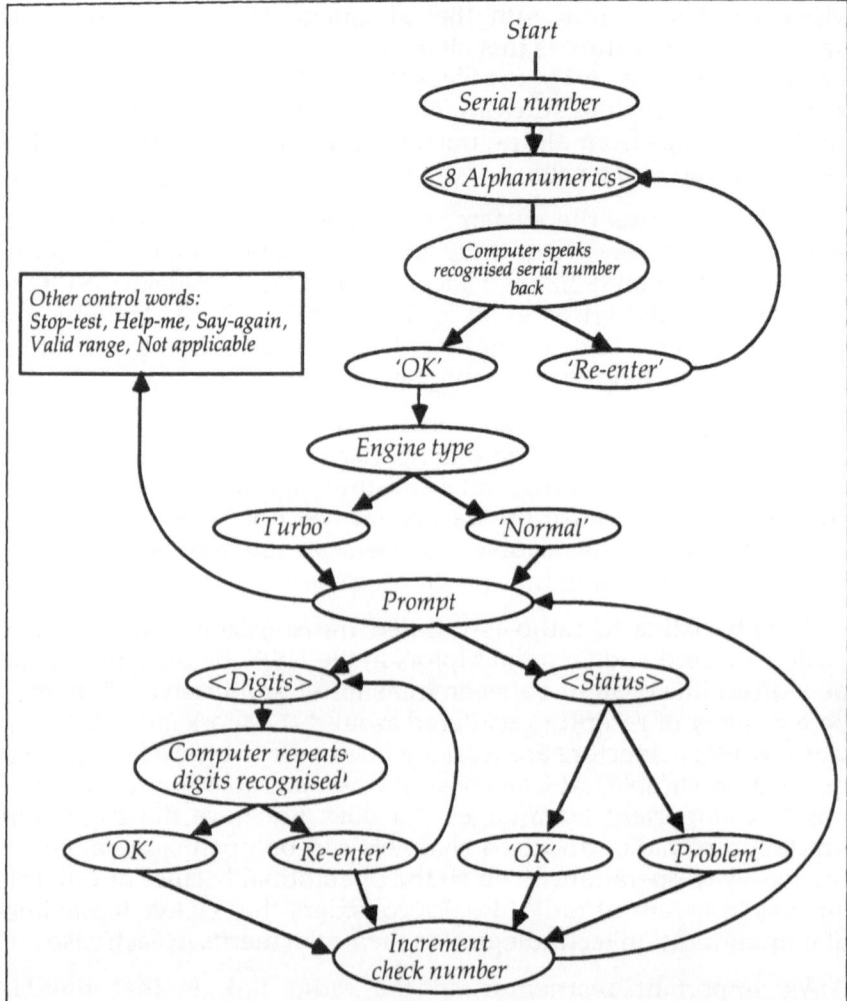

Fig. 7.b.4. Dialogue design diagram for the Caterpillar backhoe loader test station

bandwidth and very high quality transmission but was considered impracticable for single-channel operation.

The voice-activated switch is an essential element of Logica's solution. Switching on and off at the computer end to send prompts is easy but at the other end something is needed which, in effect, will switch on just before the inspector starts to speak. What actually happens is that the inspector's speech signal is split into two parts. One part goes to an activation circuit which detects the speech and switches on the radio. The other part goes through a delay line which holds back the signal until the radio is switched on and available for transmission. The switch operates in this way for each utterance spoken.

Speech input by the test inspector at Caterpillar is very different from that used at Jaguar because it consists largely of digital information – pressures, temperatures, tachometer readings and so on. It was, therefore, thought essential to play back every value with synthesised speech and to await confirmation. This has resulted in a quite interesting three-way 'conversation'. The prompts are given by a woman's voice using compressed speech output, and the numbers are repeated back by a synthesised man's voice. This adds variety to the dialogue and helps to make it seem more real to the user.

Prompts can be given at two levels – an 'expert mode' and a 'novice mode' – because when test inspectors become acquainted with the procedure they do not want to wait for lengthy prompts. So the full prompt may be 'left arm lower', or 'left arm raise', but when expert mode is called for the prompts are all reduced to one word – 'lower', 'raise' and so on.

AUSTIN ROVER

Speech recognition is being used on the production line at Austin Rover to check every finished vehicle at the end of the Rover 800 line. This imposes severe time restrictions on the inspection cycle, and four inspectors are needed, each using a separate speech recognition station, to keep pace with the line. The recognition equipment was supplied by Voice Systems International incorporating Votan speech boards, and the chief responsibility at Austin Rover for the implementation was with Kapul Gill, Principal Engineer, Robotic Assembly, who has contributed the following information.

The successful launch of Austin Rover's prestige Rover 800 series of executive cars provided an ideal opportunity to develop and introduce a new method of inspecting vehicles, which is currently being used at the Cowley, UK, assembly plant.

Highly competitive conditions in today's automotive industry have resulted in a multiplicity of trim fitment levels, engine types, gearbox variants, legislative specifications and factory-fitted options – with the result that very few consecutive vehicles on the production line are identical. Such wide variations from one vehicle to the next create difficulties for the company's inspectors, who with current techniques either have to rely on their ability to remember the different derivative checks, or have to go through the time-consuming routine of consulting large packs of data sheets. Any reports of faults are written by hand at source, and the information is not entered into a computer until some time later.

Because there is no real-time feedback of fault reporting, it may take a day before a trend in the manufacturing process is identified and corrected. The delay can prove expensive at best, and is unacceptable in a market where high quality is expected.

When Austin Rover first considered direct data entry to the computer by using speech technology, a list of requirements was drawn up to identify those features which were considered essential and those which were optional. Among the essential features were:

- Low cost.
- Use of voice input to replace manual recording with clipboards and written entries.
- Performance capable of overcoming difficulties associated with background noise, which can cause false entries and block speaker input.
- Real-time speed of response, preferably instantaneous.
- Use of a communications carrier between the operator and the computer to ensure full mobility during inspection.
- Voice prompting to enable the operator to concentrate directly on each vehicle being inspected instead of looking at a visual display of the items to be examined.

Following a review of the available technical literature on speech input/output and reviews of vendors' product offerings, Austin Rover concluded that current technology was only suitable for office or laboratory type environments, not for the demands of the shop floor. Very few companies had experience of implementing a voice system in the shop floor environment.

Several of the solutions considered were found to be impractical:

- Hard wired links between the computer and the user would inhibit the necessary mobility. In addition, hard wiring would

introduce a safety hazard, particularly with vehicles being conveyed along moving tracks.
- Infrared as a wireless communication carrier between the operator and the machine has one major drawback. The transmission path must be through either the vehicle windows or an open door, since infrared does not have penetration.
- The high cost of some speech systems precluded their use.
- From the point of view of the user it was concluded that synthesised voice prompts would after a time become very annoying, especially when long dialogues are involved.
- The use of discrete word recognisers forces on the user a totally unnatural dialogue.

As an alternative to synthesised speech output, voice compression – vocoding – generally gives a better quality and more acceptable interface. The operator has the sense of conversing with a human counterpart rather than a robotised speaking machine.

It was realised that in order to obtain the desired performance to suit the intended quality-based application, it would be necessary to undertake a period of research and development work in-house. This task was addressed by an intensive 15-month joint development programme with Voice Systems International (VSI), using the basis of an existing speech microcomputer marketed at that time.

On this basis, a working pilot scheme was developed and work began in the early part of 1986, to enable the collection of quality data.

The system in its developed form, Fig. 7.c.1, is very different from the pilot installation used at the start of the development exercise. Experience gained by both Austin Rover and VSI during the development period allowed important fine tuning of the human interface. Significant advances were made, first by standardising the vocabulary, and secondly in developing techniques during the enrolment (machine training) stage to ensure full operator interaction with the machine. It now takes only about 30 minutes of initial training before a new user can converse with the computer in a fairly confident manner. Throughout the enrolment session checks are performed on the quality of the speech patterns, primarily to identify any corrupt templates. These are simply deleted, and the appropriate words are retrained.

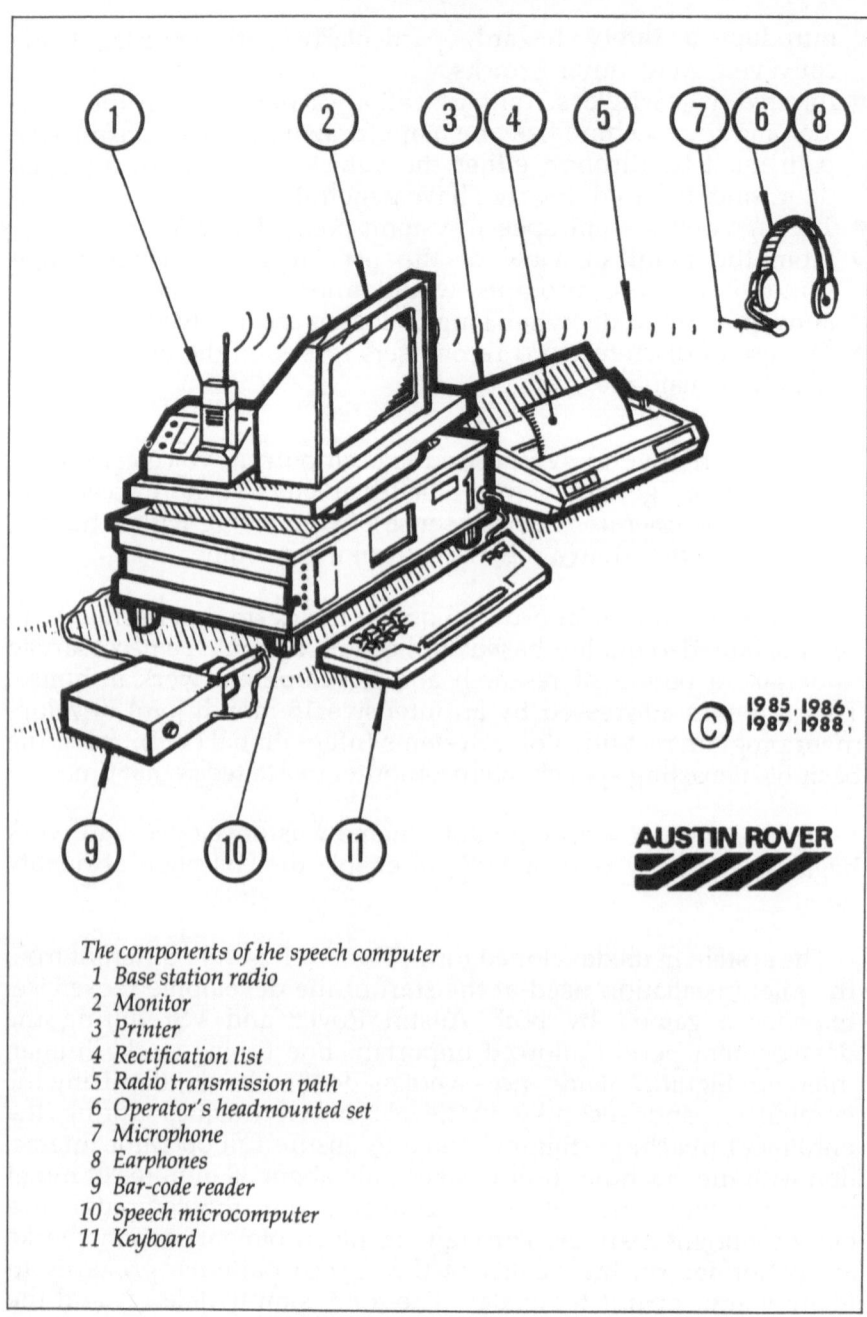

The components of the speech computer
1 Base station radio
2 Monitor
3 Printer
4 Rectification list
5 Radio transmission path
6 Operator's headmounted set
7 Microphone
8 Earphones
9 Bar-code reader
10 Speech microcomputer
11 Keyboard

Fig. 7.c.1. The speech workstation at Austin Rover

The number of retraining sessions needed varies between users. The main reasons for variations are:

- During enrolment the volume at which the user is trained is significantly different from the way he speaks on the track.
- During the first training cycle the operator tends to try to overpronounce the words in order to 'help' the machine.
- Even though the user is fairly relaxed on the track, he may be extremely tense and nervous while training.
- The natural inhibitions of speaking into a microphone have to be overcome.

The success of voice technology within Austin Rover is due to a variety of reasons. One of the most important is the high level of acceptance by the quality operators and their shop floor colleagues.

This stage of the work was seen as a development phase. During this time the system hardware was being continually upgraded and revised. Equally the software was redesigned and structured to meet the ever-increasing demands of the end user and his surrounding environment. During all the line trials operators who have had no knowledge of computers have been trained, not only to converse effectively with the computer, but also to load their own speech patterns, and most important of all to provide a collective team approach to implementing the quality assurance procedures.

There are many facets of the workstation, but three major components which are important to the success of the final engineered system are the hardware requirements; the communications carrier; and the human engineering factors.

Hardware

The microcomputer at the heart of the system is based principally on the chassis of an IBM PC–XT. It includes a basic 20-Mbyte hard disk, 512kbyte of RAM, 360kbyte floppy disk and an additional 2-Mbyte secondary hard disk. The printed circuit voice card resides in one of the auxiliary slots. The card combines both voice input in the form of continuous speech recognition and voice output as high-quality digitised compressed speech. The serial and parallel communication ports allow the computer to be interfaced with external devices such as bar-code readers, printers and visual display units, in addition to being networked with other computers.

Communications

All communication between the mobile operator and the talking computer is by natural speech in both directions, using a two-way radio voice link. The base station is computer-controlled, using a specially designed radio interface and control unit. Each operator is equipped with his own simplex Pye UHF radio, which is carried on his person, and is connected to a head-mounted set with integral earphones and a microphone.

Use of a noise-cancelling feature enables consistently high recognition accuracy to be achieved, even though the systems are being used against a compounded background noise in excess of 85dB from door shuts, horns, burglar alarms, engine testing, radios and so on.

Human engineering factors

Dialogue requirement and design. Designing the recognition dialogue is vitally important, since it is fundamental to the success of the system interface. It must incorporate certain features of the human speech process if it is to be acceptable to the end user. The key elements influencing design of a suitable dialogue are:

- *Syntax structure.* Operator input throughout the transaction is 'syntax' driven. Rules are imposed regarding the specific commands, or words which may be entered at predefined points in the interactive process. For example, the dialogue will not permit the operator to log-off during an inspection sequence until it has been completed. Restricting the vocabulary within subsets makes the response almost instantaneous and improves the overall recognition accuracy.
- *Speech transaction.* If the speech command structure is too involved and ambiguous, it will be difficult or virtually impossible for the user to remember the correct format of each command. An easily used and 'natural' dialogue between the user and the machine is necessary, with the most natural words or phrases selected for a given context.
- *Error detection and correction.* Because of the nature of speech recognition, approximately 2% of the input entries may be misrecognised. The transaction design currently supports error detection as well as provision for real-time correction.

Operator training. A training programme was developed which would ensure the competence of the operators and give them a full understanding of the technology. In addition to the procedure for training the speech patterns, it is important to take into account the conditions under which training takes place and the psychological effects of the system on the user. Such considerations have a critical bearing on the successful training of the operator's templates, and therefore on the successful and consistent operation of the speech system.

Inspection procedure

The system is best understood by describing a typical voice-driven inspection sequence as currently used for quality checks on the Rover 800/Sterling series.

User log-on and identification. An inspection shift begins with the operator identifying himself to the voice computer by a two-stage log-on procedure. The computer will allow any of 20 users to log-on. During normal operations the operator will log-on to a specific computer by speaking his inspection identification number followed by his password. At run-time, an operator identifies himself to the machine, and if it recognises a valid user, he will be given entry, otherwise access will be denied.

An operator having been identified, the computer software downloads the appropriate speech patterns into the recognition board from a local disk store. When the checks on one vehicle have been completed, the system will allow a new operator to log-on, but only after the current user has logged-off.

Vehicle identification. On arrival of the first vehicle to the inspection zone, the build card is located by the operator. This card contains information which uniquely identifies both the model derivative and other appropriate details such as trim levels and the Vehicle Identification Number (VIN). The operator uses the build card and a bar-code reader coupled to the computer to enter the relevant details. The same information can also be entered via the keyboard in an interactive question-and-answer session, to override any problems encountered with either a defective bar-code label or a hardware failure.

From the information supplied by the bar-code reader, a generic checklist for that particular model type is prepared in real time, and the first check in the sequence of 60 checks to be carried out on this

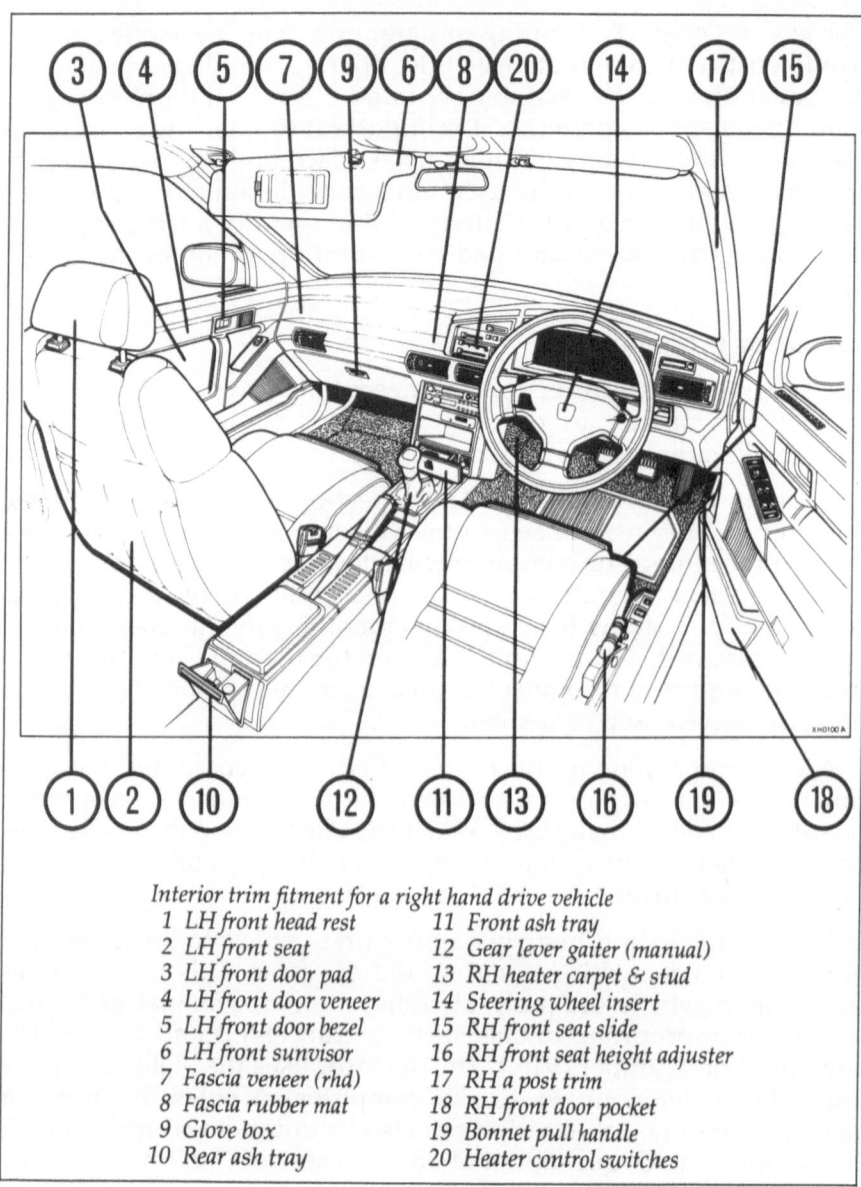

Interior trim fitment for a right hand drive vehicle

1 LH front head rest	11 Front ash tray
2 LH front seat	12 Gear lever gaiter (manual)
3 LH front door pad	13 RH heater carpet & stud
4 LH front door veneer	14 Steering wheel insert
5 LH front door bezel	15 RH front seat slide
6 LH front sunvisor	16 RH front seat height adjuster
7 Fascia veneer (rhd)	17 RH a post trim
8 Fascia rubber mat	18 RH front door pocket
9 Glove box	19 Bonnet pull handle
10 Rear ash tray	20 Heater control switches

Fig. 7.c.2. Some of the interior trim items checked using voice input

car is prompted to the operator. Some of the interior trim items prompted by the computer are indicated in Fig. 7.c.2 and an inspector is shown at work in this area in Fig. 7.c.3. Fig. 7.c.4 shows

Fig. 7.c.3. Inspector carrying out quality assurance check on interior trim fitment

Fig. 7.c.4. Inspector carrying out an under-bonnet check

an inspector using the talking computer system while he is carrying out an under-bonnet quality inspection. The hardware he carries is clearly visible in this picture.

Operator dialogue. The inspection task consists primarily of a dialogue of prompt and response by natural speech in both directions. The operator talks to his own computer through the two-way radio link, and the computer prompts the user through his headset.

There are various types of verbal prompting and response available to the operator. Principally the check can be either passed or failed, but there are also many other commands at the operator's disposal, as shown in the dialogue structure in Fig. 7.c.5. For example:

- The user is able to repeat the previous check after the computer has proceeded to the next check in the sequence.
- He can request further information about the specification relevant to a check by asking 'repeat check', in which case a fuller explanation is transmitted.

If, in the normal case, the item is passed, the prompt for the next check in the sequence is heard. Should the item be failed, the operator can choose from several predefined lists of fault descriptions such as 'damaged', 'security', 'connection', 'profile' and so on – words describing a reason or combination of reasons for failing the item being inspected. As each fault description is spoken by the operator, it is echoed back by the computer. If the word echoed back is not what was said, the dialogue design allows easy correction simply by cancelling the last input and entering the correct word. After all the faults relating to the current prompt have been entered, the operator says 'all-finished', and the computer describes the next feature to be inspected. When the last check has been completed, the computer tells the operator that the full list of checks has been carried out.

At this stage the results are written to disk for subsequent statistical analysis, and a hard copy in the form of a rectification list is printed out. The computer then asks the operator to initiate inspection of the next vehicle with the prompt 'bar code'.

The rectification sheet printed out for each vehicle can either be a complete list of the checks performed together with all the

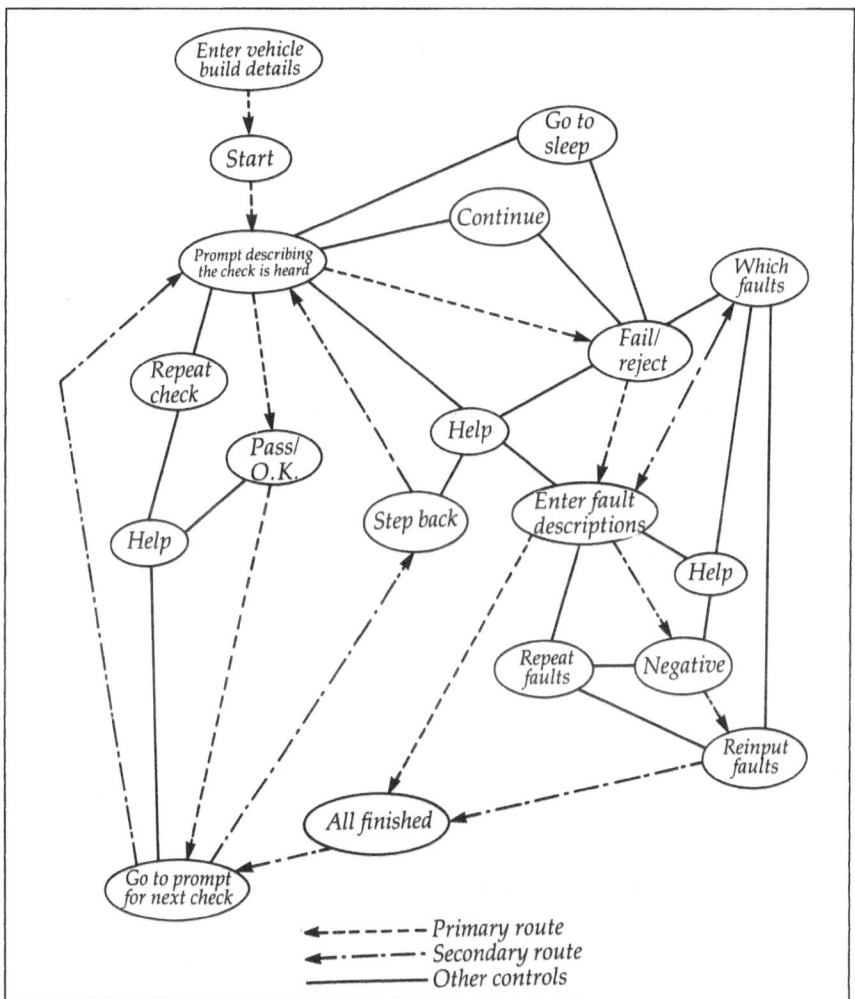

Fig. 7.c.5. Dialogue design structure for Austin Rover quality assurance

appropriate responses obtained, or simply (as used currently) a list of the failed checks cross-referenced with their fault descriptions.

Statistical analysis software. All the information gathered by the four stand-alone voice computers during inspection is recorded on disk and used for statistical analysis and management reporting. The fault statistics reports provide a printed or displayed listing of the 10

– or optionally up to 20 – checks that were failed most often in a specified period of inspection, together with the three (or optionally up to 10) faults on which those checks were failed most often. Typical statistical information available is:

- Success and failure rates.
- Most common faults.
- Percentage of vehicles fault-free at final inspection.
- Inspection zone performance rates.

Benefits

The most important benefit from the speech interface is the real-time nature of the data capture process. As against conventional methods where the information is analysed hours or even days after the event, the almost instantaneous interrogation of the system offers a real-time picture of both the inspection and the manufacturing process, allowing rapid feedback to the production zones for rectification. The information can also be relayed to product designs, production control and quality so that the problems can be isolated and immediate effective action can be initiated as a coordinated effort.

Among other important benefits are the following:

- Consistent checking of a predetermined list ensures consistently high inspection standards.
- Inspection is totally controlled by the speech computer. The inspector does not need to remember the checks for every model derivative.
- The disciplined nature of the system ensures that no cars can be released to the next stage until the procedure has been completed.
- Recording and transcription errors are eliminated.
- Absolute integrity of data is maintained throughout the system.
- The system can adapt to incorporating new checks as necessary to satisfy company quality requirements.

Refinement work is currently in progress to develop a fully integrated quality assurance system.

ROLLS-ROYCE

Ultrasonic inspection of turbine blades at the Precision Casting Facility of Rolls-Royce at Derby is being achieved very much faster, and computer capture of the information is being obtained at the same time, with the help of a speech system installed by Voice Systems International using a Votan speech card in an IBM PC. The pilot application has been so successful that the company is now planning to apply speech systems to a number of other tasks.

The operation, Fig. 7.d.1, is an ideal one for speech in that the inspector's eyes and hands are fully occupied. Without the voice input she had to break off inspection to write each result on a report sheet. Any analysis of the reports then had to be done later by hand.

The quality requirements for aero engines are extremely strict, and every gas turbine blade must be inspected in detail and a record kept of the inspection. The task involved in the speech application is measuring the wall thickness of the blades. Blades are investment-cast in nickel-based alloy and are hollow, with a number of passages for air cooling, separated by webs, and leaving as little as 0.014in. of metal under the blade surface. This thickness is critical for the successful performance of the blade, and must be maintained within the design specification. As the cross sections of Fig. 7.d.2 show,

Fig. 7.d.1. Ultrasonic inspection of turbine blades requires constant attention to the oscilloscope screen (below the computer monitor) while manipulating the blade

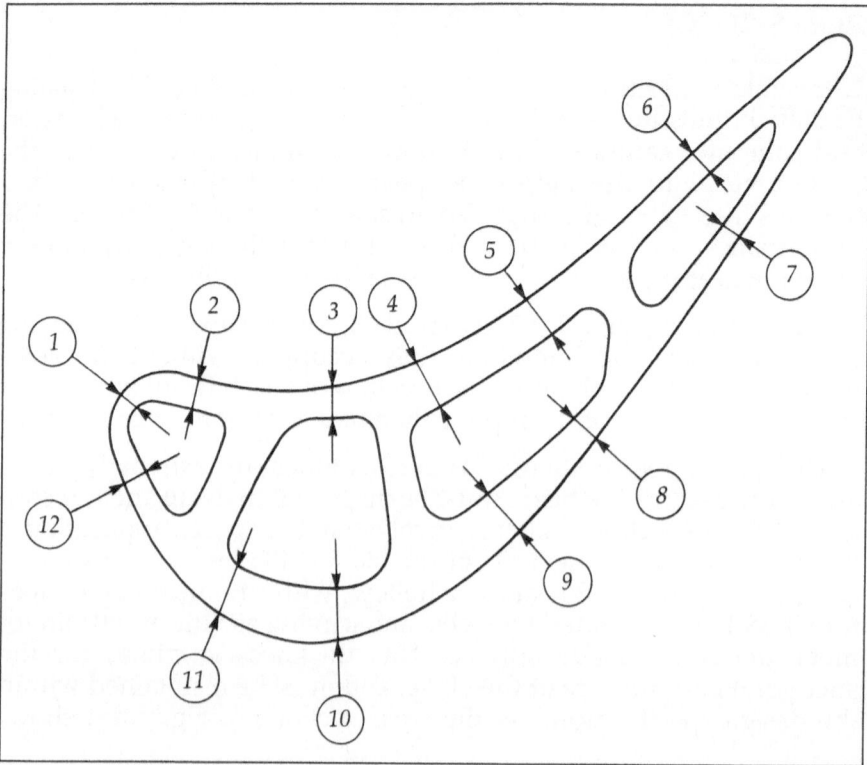

Fig. 7.d.2. Stylised example of an aerofoil section indicating points of measurement

there may be several places around the chord of the blade where the wall is thin, but there is nothing on the outside of the blade to indicate this.

Ultrasonic inspection is used to measure the thickness of the walls. This employs a high-frequency sound wave from a transducer, which strikes the surface of the metal, passes through it and is reflected off the back surface. Reflections are picked up from the front and back surfaces, and the time taken for the sound wave to travel through the metal is shown by two 'blips' in a line on an oscilloscope screen. A contact ultrasonic transducer would cover too large an area – readings would not be acceptable because of the small and intricate passageways inside the blade which must be avoided for these ultrasonic measurements. The probes are of the focused beam type. Pulses are focused on a small area of the blade surface through a plastic cone through which water, used as the transmission medium, flows on to the surface of the blade.

The inspector's task is first to mark with a crayon, using a template, a series of lines around the chord of the blade at the sections to be inspected; then to hold the blade in contact with the tip of the cone, and to move it along each marked line in turn around the blade. At the same time she must watch the oscilloscope screen and note the value of the wall thickness. To help her interpret the oscilloscope readings she has in front of her a cross-section drawing of the blade, Fig. 7.d.3, so that she can see and count the cavities being detected. This process card also carries the part number, details of all the areas to be checked, and the tolerances to be applied.

Most blades are simply inspected to ensure that the wall thicknesses are within the required tolerances. In some cases, though, for example where a die has been changed or redesigned, the engineers require that the actual thickness be recorded for a batch of blades, and it is this task which, without speech recognition, is extremely laborious.

Before the speech system was installed, the inspector had to release her grip on the blade to write down the values on a report sheet, then pick up the blade and find where the last reading had been taken before going on to the next one. With the new system she simply speaks the dimension into the microphone and moves the blade on to the next position, completing the operation with

Fig. 7.d.3. Inspector uses cross-section drawings to identify the numbers and locations of cavities to be detected

perhaps 50 readings without putting down the blade. The procedure is now several times faster than it was with the previous method.

In conjunction with Rolls-Royce, the company which supplies the ultrasonic equipment, Meccasonic, also has been involved in the development of a digital thickness readout system which could be interfaced directly with a computer output, but unfortunately this only functions effectively with thicknesses greater than 0.02in. This would severely limit the number of blades for which the method would be usable, leaving many others for which readings would have to be taken by the inspector. The speech input method supplies a reliable procedure which can be used for all types of turbine blades.

Before entering the ultrasonic thickness readings, the inspector uses the microcomputer keyboard at the beginning of the operation, first to enter her own identity number to call up the vocabulary which she has taught to the system, then to enter basic details about the particular turbine blade from the paperwork that comes with it and from the blade itself. The details include a trial number which has been supplied by the quality or methods engineers who have called for this detailed inspection; the part number; the number of blades in the batch; the number of sections of the blade to be checked; the die number on the casting which identifies the particular die used; and the serial number identifying the individual blade.

The number of readings to be taken having been enterred into the PC, it expects that number of readings to be entered and shows a list of reading numbers on its screen. As the inspector speaks each dimension, for example 'zero two four', the figures appear on the screen alongside the reading number as confirmation that the message has been received correctly. The inspector has a number of additional commands, listed in Fig. 7.d.4, which allow her to control the process. For example, if on reviewing the data certain figures appear inconsistent, she can 'jump' to a particular reading number, check the measurement and enter a new value if necessary. 'Hibernate' allows her to disable the recognition function so that she can speak to somebody else or leave the workstation. 'Wake up' reactivates the recogniser.

When the equipment was discussed with one of the inspectors, she said that she was quite pleased with it. Its recognition accuracy

HELP
GOBACK
JUMP
THRESHOLD
HIBERNATE
ABORT
ALLDONE
CANCEL
YES
NO
WAKE UP
Digits 0 through 9

Fig. 7.d.4. Command vocabulary of Rolls-Royce inspection system

was good except for a tendency for more errors towards the end of the day as her voice became more tired. If she had a cold she usually found it necessary to retrain the vocabulary, which with only 21 words takes only a few moments. The work itself requires considerable dexterity and concentration, but some variety is provided with the keyboard entry between inspections. Like others starting to use speech recognition, she had found it difficult at first because of self-consciousness and a tendency to speak unnaturally. It took her about 5–6 weeks to reach complete self-confidence with the system.

Background noise in the inspection area is surprisingly high, but mostly does not affect the recogniser. In fact the equipment has even been used successfully in the fettling area where the noise level is such that it is difficult to carry on a conversation.

As often happens, the biggest benefit to Rolls-Royce from the system has come not from the inspection operation itself but from the tremendous time saving that has resulted from direct computer entry of the data. Because the results go straight on to floppy diskette, the disks can be taken away and analysed directly with specially written statistical software, which is still in process of refinement. Even with the software in its present incomplete state, though, analysis of a batch of blades which used to take about one and a half days of an engineer's time can now be completed in about 15 minutes, including the production of histograms and details of standard deviations and ranges. On the shop floor, the application of the voice system has resulted in a threefold mean improvement in throughput, and on certain components as much as fivefold.

The system first went on the shop floor in May 1987 and was fully implemented in September when the analysis software had been completed. In March 1988 Rolls-Royce was planning to buy at least one more system for use on other tasks of a similar nature. This would also ease the pressure on the existing system, which has been so much in demand that it has been difficult to spare the time for new inspectors to become familiar with it. It generally takes about two weeks before an inspector is sufficiently skilled with the equipment to use it under ordinary working conditions.

No difficulties have arisen with the dialogue design, which is quite simple. There have been no problems with the hardware, but there has been continuing experimentation with alternative microphones which VSI have supplied in a search for a headset which gives good recognition performance, is comfortable to wear all day and can be accommodated to the inspectors' hair styles – all of which are important practical considerations.

FORD DISTRIBUTION CENTRE

Running successfully since 1983 has been a speech recognition system in the Ford Motor Company's parts distribution centre at Cologne. The centre handles about 65 000 different parts and distributes them to 1200 domestic dealers, about 15 European Ford locations, and Ford companies and dealers worldwide. About 40–50 lorries and railway wagons are loaded daily. Before the speech system was installed, all the shipping data had to be written down on a number of different papers and later had to be keypunched by a typist.

The speech system was developed by Computer Gesellschaft Konstanz, a Siemens subsidiary, and includes eight voice workstations in the shipping area entering data for containers and packages leaving the warehouse. As a result of introducing the speech system:

- On-line processing of data provides faster and more accurate handling and avoids double work.
- Data can be entered while physically handling material and while physically moving around the working area.
- Comparatively little expense is incurred in training personnel.

The system uses a radio microphone input and feedback to allow the operator freedom of movement.

In the picking area materials are picked and put into boxes or containers, and forwarded to the shipping area. Order processing in the shipping area is the last step fully controlled by Ford, because the transporting is done by independent companies. It is, therefore, particularly important that information should be accurate, as it is used for invoicing, for customs papers if the consignment is going abroad, and for matching against any claims.

People working in the shipping area have a number of tasks. They move the material, with or without mechanical help; they record invoice data; they load lorries and record data in the form of a loading list for a lorry. In addition, they combine several packages for one dealer into a single package if possible, and they record the final package details on the invoice. The work involved is, therefore, a combination of manual handling and data acquisition. Before the computerised system was introduced, the data acquisition involved writing different lists containing partly the same information.

The shipping computer system gathers the information and prints the loading list for every lorry leaving the warehouse. It is linked to the distribution centre's mainframe computer, and provides statistical and other information for the foremen, besides maintaining the speech-based dialogue with the operators, as in Fig. 7.e.1.

Every day, before the warehouse operation starts, order details are sent from the mainframe to the shipping computer. They are stored and put in the correct format for on-line processing during the day. Invoice details include dealer, destination, estimated volume and weight, and for every expected package a unique package number is generated, which is used for identification in the picking area as well as later in the shipping area. After every shift a post-processing operation takes place in which lists are produced of finished invoices, statistics about shipped material, and files which are transmitted back to the mainframe.

The shipping computer is a Siemens R30 minicomputer with two fixed and two floppy disk drives, 1Mbyte of storage, and peripherals. It is used to control the user dialogue with the workstations; to maintain the link with the IBM mainframe; to store, control and update data on the disk drives following voice or keyboard transactions; to perform data security and, if necessary, recovery; to control the user dialogue via keyboard terminals for certain

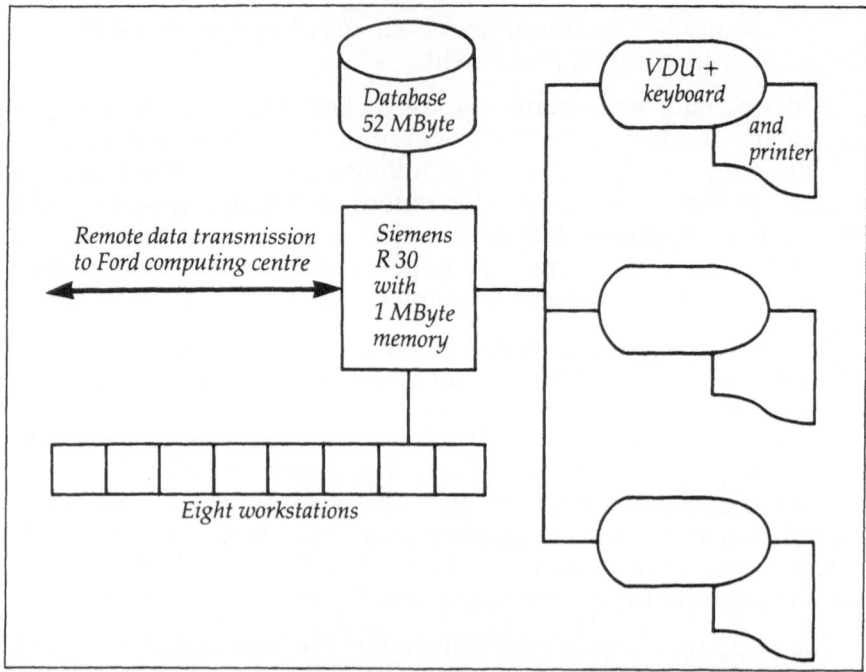

Fig. e.1. Configuration of the computer system with links to the mainframe computer and to speech workstations

functions; to carry out the processing before and after each shift, and to control initial training for users.

Each of the eight workstations is run by one person, assisted at times of peak load by helpers who only carry out handling tasks. Handling consists of taking a container off the towveyor system, pushing it to the floor scales, possibly adding parcels for the same dealer, closing the container with a plastic cover if necessary, and pushing it to the appropriate loading dock. In a second step the containers are brought on to the lorry. Most of the work is done with self-rolling containers but, depending on the circumstances, fork-lift trucks and other mechanical aids are used.

The operator carries a lightweight headset consisting of a noise cancelling microphone and earphones connected to a lightweight FM transmitting/receiving system operating in a bandwidth of 6–8kHz. Signals are received by strategically located antennae and brought to the central fixed receivers. From these the analogue signal is passed into the voice recognition unit which carries out the

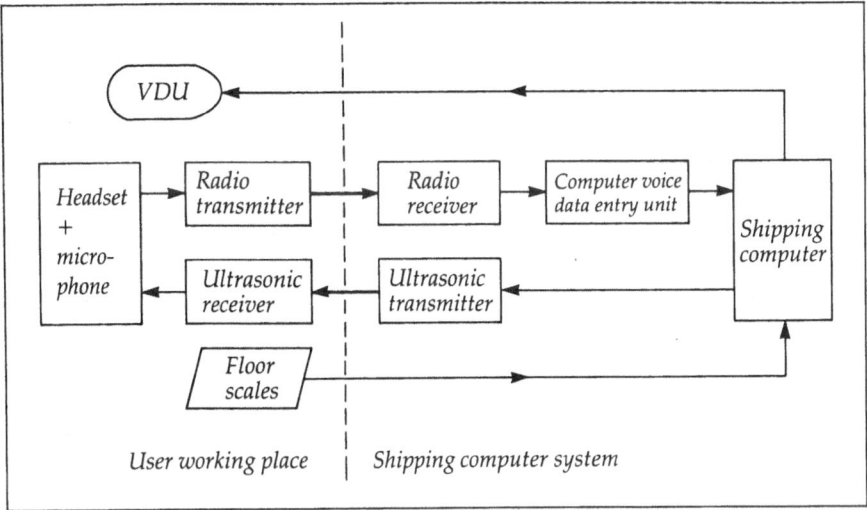

Fig. 7.e.2. Communication with the operator is via radio and ultrasonic links and large VDU displays

template matching and sends the recognised word in ASCII characters to the shipping computer.

Feedback to the operator takes two forms (Fig. 7.e.2). The audio feedback is not by radio but by an ultrasonic link which signals an error by triggering one or two beeps in the user's earphones, depending on whether it was an application error or a wrong word. Ultrasonics was adopted because the use of radio is legally restricted and most available channels had already been used for the voice input.

Feedback to the operator is also provided by large-screen VDUs, which are installed in sufficient numbers to be readable wherever their information is needed. Displays show all recognised words, guide the user and give all necessary information for carrying out the task.

There is one other on-line link to the computer – the floor scales (Fig. 7.e.3). These have a direct connection to the computer, and the weight reading is transmitted directly on receipt of a spoken command from the operator.

Packages arriving in the shipping area are handled in two steps:

- Consolidation of different parcels into one package for one dealer and registration of package data, consisting of package number,

Fig. 7.e.3. Container load is pushed on to the floor scales, and the weight is automatically transferred to the computer on the operator's voice command

packing type, weight, measurements, identification of invoice number, and decision if invoice is completed.
• Capture of loading data, registration of package for a lorry or wagon with a given destination.

Both tasks are carried out by a dialogue with the computer. The operator reads the package number from a label on it, and this identifies the package to the computer which supplies invoice and destination details. The operator specifies the type of package, and the weight of the package on the floor scales is read automatically by the computer on a spoken instruction from the operator. The amount of information to be entered by the operator is thus kept to the minimum. Progress of the transactions is displayed on the VDUs.

The input is given a formal and a plausibility check at three levels. The formal check is, for example, whether an alphabetic character or a digit is required. There is a syntactical check that a digit is correct, and a plausibility check that the invoice number is actually stored in the system.

Keyboard input is required only for setting up such information as destination, lorry number plate and dock number; changing or deleting invoices; printing statistics; and initial voice training of the vocabulary of 60 words.

The Ford Cologne Distribution Warehouse is remarkable in that it must be one of the longest-running implementations of speech recognition in Europe.

8 THE FUTURE

SO MANY exaggerated claims have been made in the past for speech recognition systems that the reality, for several years to come, will probably appear commonplace. No dramatic advances in recognition systems are likely to appear before the 1990s. What we shall more probably see are steady improvements in accuracy and in vocabulary size, with falling prices and imaginative ideas bringing a substantial growth in applications in certain areas.

Much of the growth in applications will result from a growing understanding of what is and is not currently possible with speech recognition, and a tailoring of systems to make the most of what can be done reliably. At the same time, development work on recognition algorithms, together with the growing capacity and speed of computers and falling costs, will bring advances in continuous speech recognition.

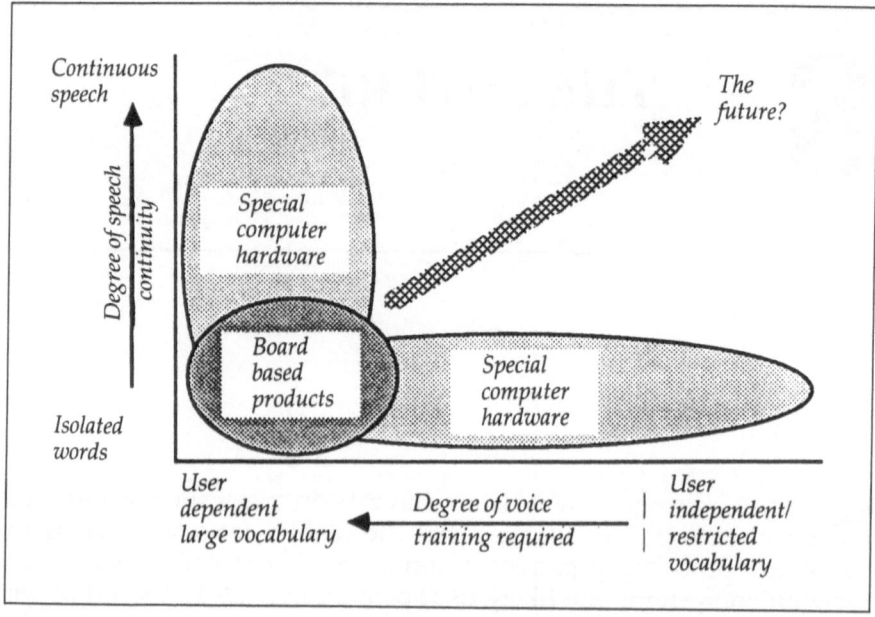

Fig. 8.1 Improvements in speech recognition systems will be in the direction of accepting greater continuity of speech with larger vocabularies and greater speaker independence

One direction of study at present is to look at *allophones*, the building bricks from which phonemes are constructed, and which are modified when words are joined together by 'co-articulation'. There are some rules governing co-articulation, but it is affected by many considerations such as speed of talking and local accents, and interpretation of spoken input is going to require an artificial intelligence approach.

Speaker-independent systems can be expected to become more robust and to acquire larger vocabularies, and will move towards a very limited capability for recognising connected words over the telephone. This in turn will have an effect on the growth of telephone voice response systems in Europe, where DTMF tele-phones are not as widely used as in the USA. They will also increasingly be available as an option to touch-tone input in the USA. In the longer term, speaker-independent vocabularies of a few hundred words will be possible.

TELEPHONE VOICE INPUT AND RESPONSE

A sign of the trend towards the availability of speaker-independent, relatively large vocabularies is the experimental telephone banking system being run on a pilot scale in 1988 by the Royal Bank of Scotland in association with British Telecom. Customers are able to phone the bank's computer and obtain details of their accounts, arrange transfers between accounts and pay bills. They do all this using entirely spoken dialogue with a mixture of speaker-dependent and speaker-independent recognition, and they need no equipment other than a telephone.

When a customer calls the bank, the computer voice, which is concatenated coded speech at 16kbits/s, asks for the customer's five-digit personal code and password. When this is given, the computer refers to a previously stored collection of voice prints from that customer and asks the customer to speak one of those words, which is compared with the voice print. It also carries out other security checks to verify the identity of the customer. Speaker verification is an essential element in this system – it would not be enough simply to rely on an identity code and a password.

Having validated the identity of the caller, the recognition system switches back to speaker-independent recognition and customers are able to ask for balance of accounts; obtain details of the last six transactions; request a cheque book or statement; pay bills or transfer money between accounts – either internal or to settle other accounts such as electricity, gas, telephone or Access.

The entire interface unit, including voice validation, speaker-independent recognition, low bit rate coding for voice generation and telephone interface, is contained on one card designed to slot into an IBM PC or compatible. It was developed at British Telecom's Research Laboratories.

Because the system now offers speaker-independent recognition over a telephone line, it opens up opportunities for use with any kind of telephone database. Most telephone access databases require some form of security, if only to protect them from malicious interference, so the full system used by the Royal Bank of Scotland would be appropriate.

A more ambitious project due for completion in 1988 is VODIS – Voice Operated Database Inquiry System – running under the UK's Alvey Programme with Logica, British Telecom and Cambridge

University in partnership. The aim is to go beyond the capability of present-day voice database inquiry systems in having a more conversational speech interface with intelligent dialogue control, for use by members of the public. This requires attention to several matters, including human factors in the design of the dialogue; increased capability of speaker-independent recognition; and increased naturalness of text to speech synthesis, both in intonation and quality.

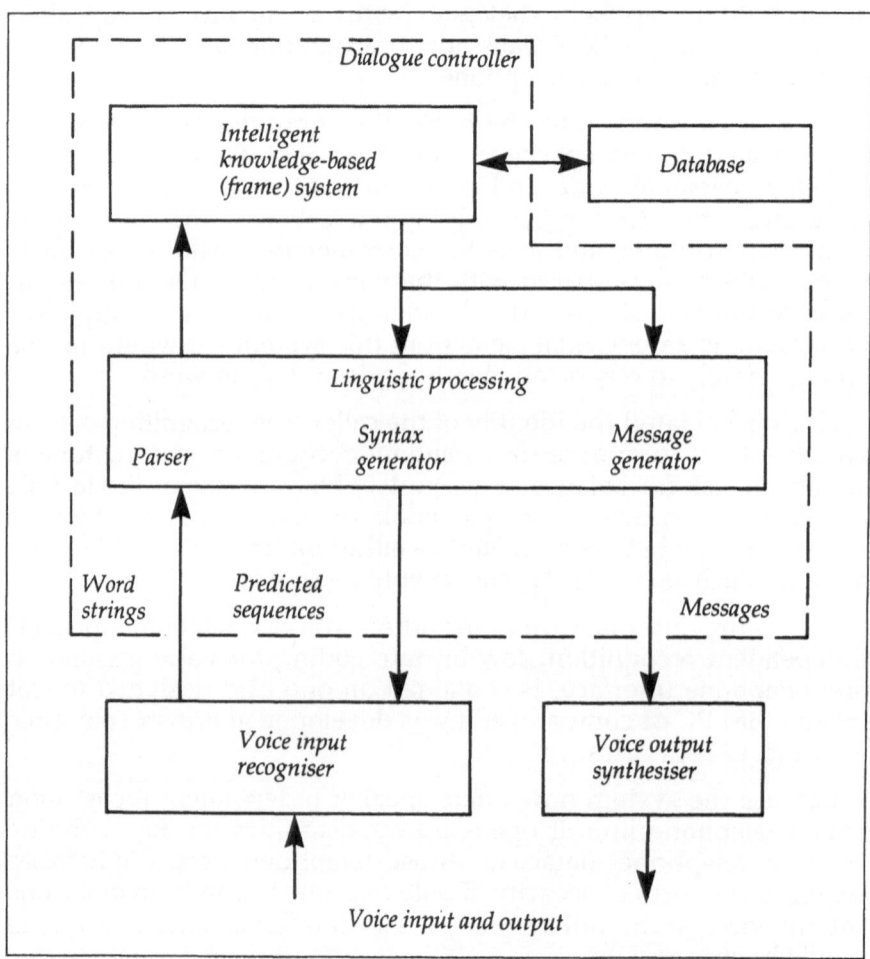

Fig. 8.2. Diagram of VODIS (Voice Operated Database inquiry System) (A collaborative Alvey project on man-machine interface involving British Telecom, Logica and Cambridge University)

The system used as a starting point a frame definition language called UFL developed by Dr S. J. Young of Cambridge University, Logica's Logos speech recognition system and a text to speech synthesis system developed by British Telecom (Fig. 8.2).

Present telephone inquiry systems are highly structured, with the system asking a series of questions and expecting a limited number of answers. VODIS is intended to give much more freedom to the caller in structuring the conversation. So, in a railway timetable enquiry application, the system would start by announcing itself and asking a question, such as 'What train do you require?'. There are many possible responses, but they will probably include a destination or a time (or something like 'tomorrow morning') or both. The system must identify the key words like times of day, station names, 'from' and 'to', and remember the information it has in asking the next question. It should be able to cope with users who answer questions which were not put – 'Where do you want to leave from?', 'I want to leave about nine o'clock'. Young's frame language provides a method for managing dialogues like this, and a typical dialogue between a user and VODIS is shown in Fig. 8.3 alongside a transcript of a live telephone inquiry.

LANGUAGE TRANSLATION

The technique of identifying key words in a speaker's natural flow of language is being used in other projects currently in progress at the British Telecom Laboratories – perhaps most impressively in the work on language translation. This work began with a study of automatic translation of text, but quickly moved on to using speech input and output. A system has been demonstrated with an English and a French speaker, in which the English speaker makes a reservation for a hotel room. The messages are automatically translated and spoken in French, and responses in French are in turn translated into English for the person making the booking. There is considerable freedom in the structuring of the requests and the answers, and various associated requests, like a room with a bath, distance from the station and so on can be accommodated.

The system, which is illustrated in Fig. 8.4 is based on about 400 corresponding phrases in English, French, German, Swedish, Spanish and Japanese, and because of its structure could be used for translation between any of these languages. At present, it is limited

VODIS:	*Which train do you require?*	
User:	I want to go to London this morning.	
VODIS:	*Sorry, where do you want to go?*	
User:	London.	
VODIS:	*From Manchester to London?*	
User:	No.	*British Rail, can I help you?*
VODIS:	*Where from?*	Hello, could you tell me the
User:	Stockport.	times of the trains from
VODIS:	*To London from Stockport?*	Ipswich to Peterborough
User:	Yes.	after 4:30 this evening?
VODIS:	*You want to go between 6 o'clock and 12*	*After 4:30?*
	o'clock?	Yes.
User:	Yes.	*Just hang on a minute.*
VODIS:	*The 7:35 from Stockport arrives in London*	Thank you.
	at 10:05. Do you require further	6:35.
	information?	6:35?
User:	I need to arrive before 10:00.	*Yes, direct train gets there 8:28.*
VODIS:	*The 7 o'clock gets in at 9:37.*	8:28. That's lovely. Thanks
	Is that alright?	ever so much.
User:	When's the one before that?	*Thank you.*
VODIS:	*The 6:30 gets in at 9:06.*	'Bye.
	Is that O.K.?	'Bye.
User:	Yes.	
VODIS:	*Thank you, goodbye.*	

Fig. 8.3. The difference between computer and human dialogue. Left: dialogue of a user interacting with VODIS. Right: transcript of a naturally occurring enquiry

to situations with a very restricted range of possible key words, but it is quite possible that it will be in use on an experimental basis by 1990, perhaps between international telephone operators – who have a quite small number of different conversations. Further possibilities where a limited phrase book would be adequate might be for engineers working on the Channel Tunnel, and for traffic controllers once it is open. There are increasing numbers of situations where people need to communicate across language barriers with the use of a small specialised or technical vocabulary, and it is here that speech language translation has a good prospect of success in the near future.

IMPROVEMENTS IN RECOGNISERS

Industrial and other situations where background noise is a problem will benefit from increasing noise robustness of speech recognition

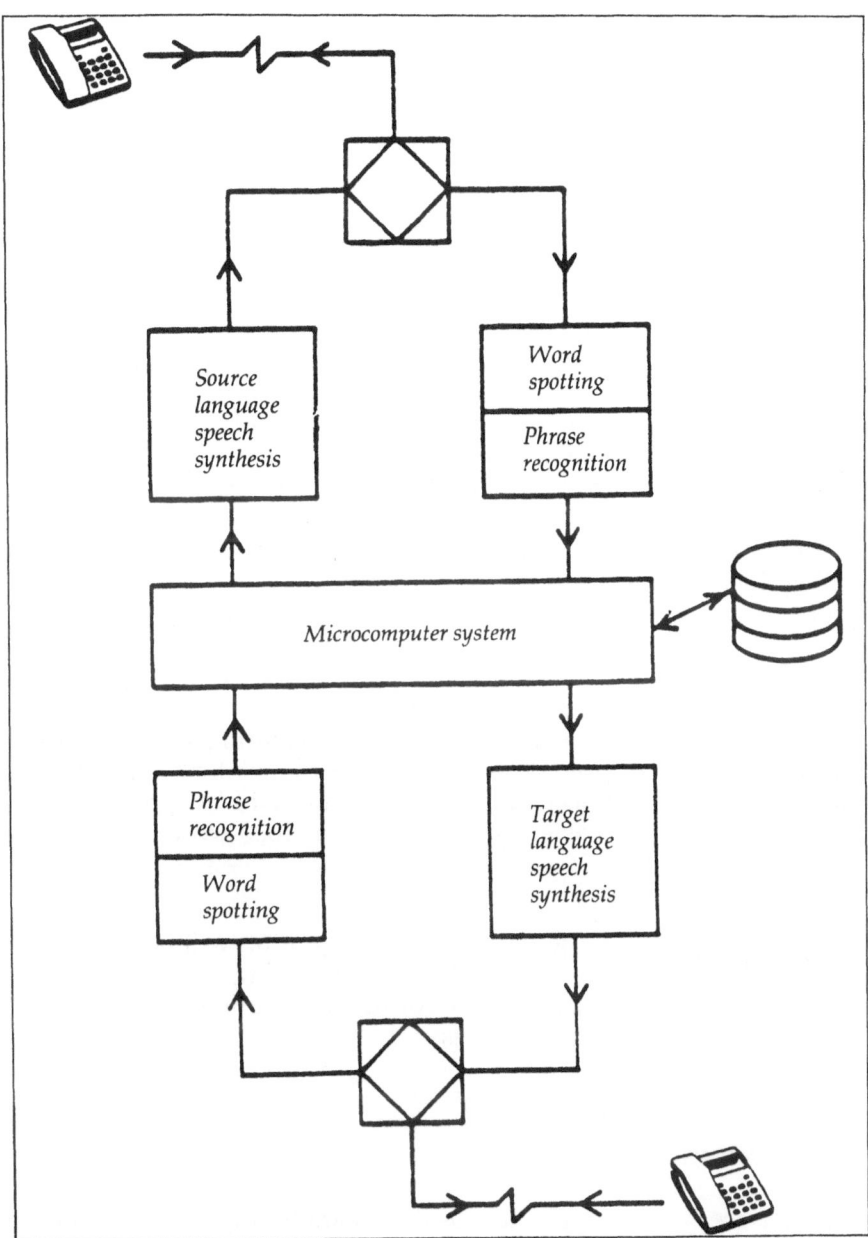

Fig. 8.4 Principle of a translating telephone system (Courtesy of British Telecom Research Laboratories)

algorithms as a result of research aimed at coping with extreme military and aerospace environmental noise.

Training of recognisers becomes more and more troublesome as the vocabulary size increases, and there is work in progress aimed at limiting the demand on the speaker to repeating a relatively short passage containing all the phonemes or allophones which will be used in the much larger vocabulary. From this information the system will then construct a synthetic set of templates for all the words in the vocabulary.

Facilities such as this will be an essential requirement of 'voice typewriter' speech to text systems when they eventually appear on the market. A major stumbling block with such systems is the need to distinguish words in a very large active vocabulary with very little opportunity for subdivision. Grammatical and pragmatic rules can be applied, at least in theory, to reduce the number of words likely to follow a particular word, but if, for example, the next word can be a noun, or begins a sentence, the choice is very wide. Some ambiguities can be resolved by the context of the sentence, and the problem is somewhat eased if alternatives can be presented immediately on-screen to the person dictating the message so that corrections can be made as dictation proceeds. One of the companies that has been working for some years to develop a voice typewriter, Kurzweil, has found a partial solution by going for a specialist medical application area where the sentences are highly structured and the vocabulary consists largely of lengthy technical words, making it easier to construct a 1000-word vocabulary without too much danger of misrecognition. Even so, the system employs isolated-word recognition.

Major progress in large-vocabulary recognition will have to await further development in speech understanding as a branch of artificial intelligence, taking account of phonetics, linguistics, semantics, syntax and pragmatics.

RECOGNISER ASSESSMENT

Another area which will become increasingly important as speech recognition becomes more widely used is setting standards for performance assessment of recognisers. Many suppliers of equipment give a figure for its recognition accuracy, usually in the region of 98%–99%, or they quote an error rate, which is the corresponding percentage subtracted from 100. Sometimes they give details of the test vocabulary with which the performance was obtained. Unfortunately, however, there is as yet no commonly agreed vocabulary or test procedure against which performance can be measured and compared.

The lack of a performance standard is not simply a matter of manufacturers failing to agree. There are very serious difficulties in the way of determining a satisfactory standard because it is not easy to distinguish errors attributable to the equipment, such as the recognition algorithm or the type of microphone used, from external variations. Many variables can affect the performance of the system. These include the choice of the vocabulary to be used, the training strategy adopted, and the inconsistencies which speakers introduce in their use of the system because of the way they have been trained, their motivation for the task, fatigue, background noise and any other tasks being carried out at the same time.

Research on recogniser assessment aims both to study performance to help in the development of recognition algorithms and to give users a means of deciding which is the best recogniser to choose for a particular application.

Currently there are two main ways of carrying out an assessment:

- Conducting field trials.
- Using prerecorded speech databases containing a representative selection of likely speech 'tokens'.

Both methods use the recogniser on test to process incoming speech directly, and both methods can provide a single number which will be a measure of the recogniser's performance. Field trials are probably more accurate, but are expensive and are elaborate to control. Prerecorded databases, once recorded, are simpler to use, but may be less representative of field conditions. They also limit the investigator's choice of experiments to the types of speech contained in the database.

Research is now being undertaken to extend the capabilities of the prerecorded database so that it is more representative of speaking conditions in the natural world. This approach, called Recogniser Sensitivity Analysis (RSA) is being developed by Logica within the Alvey-sponsored Speech Technology Assessment project. It is based on a database which is constructed with a carefully controlled range of variability in many speech parameters. These parameters are selected to encompass both the entire range of human speech variability and the effects of the environment on the speaker. They include speaking rate, speaker consistency and phonetic 'confusibility'.

The output from the RSA experiments is not simply a single performance figure, but rather a numerical picture of the recogniser's overall performance. Any particular environment can then be investigated and the data used to predict the recogniser's performance for that environment. The essential difference between the RSA approach and other prerecorded database approaches is that RSA can be applied to any particular environment and any given application, because what it attempts to model is not applications, but speakers and environments.

Because its distinctive approach does not limit it to any particular application, RSA is also being extended to address the problems of multi-lingual recogniser assessment through the Esprit Speech Assessment Methodologies project. This project is concentrating on the differences between a number of European languages, so that they may be included in the overall RSA.

It is hoped that RSA will eventually provide a standard for recognisers similar to the way benchmarks are used for comparative studies of computer performance.

TEXT TO SPEECH

As with speech recognition, advances in speech understanding will lead to major improvements in the quality of spoken output – in intonation and timing – making the voice sound more natural. There will be greater opportunities for modifying the voice. It is already possible to switch between a number of standard voices and to modify these voices, but better understanding of the ways in which pronunciation is modified, together with lower-cost and more powerful computing capacity, will lead to opportunities for generating regional accents.

Alongside developments in text to text language translation, text to speech synthesis will allow spoken as well as printed output from foreign language texts.

SYNTHESIS BY ANALYSIS

Speech synthesis, which has been on something of a plateau recently, is set for further advances, both in areas of application and in technology resulting from advances in speech coding.

A reappearance of in-car speech synthesis can be expected in the near future with the introduction of road guidance systems. A demonstration system that has gone on trial in London in 1988 under the auspices of the Ministry of Transport uses both a visual display and an optional synthetic speech output. The system, which has been under development by the Transport and Road Research Laboratory and is expected to go into service in the early 1990s, employs a network of beacons using infrared transmission to and from vehicles fitted with the guidance equipment. The beacons will be linked to a control centre so that traffic flow information can be updated continuously and drivers can obtain guidance on getting to required destinations and also will be given a route avoiding currently congested areas or traffic accidents. Between beacons, a dead reckoning method using the vehicle odometer and a compass will establish the position of the vehicle.

The system incorporates a very simple visual pictogram display associated with a 'bleep' sound to give a warning when it is updated. It will indicate when the driver must take a turning, and will show a bar which shortens as the vehicle approaches the turning. The speech synthesis system, on which Texas Instruments is working, gives the same information audibly, using a vocabulary of between 40 and 50 words. There is much to recommend speech output in an application like this because a screen display in the vehicle demands some deflection of attention from the road, however momentary. The guidance system operates in real time, telling the driver what to do next, with instructions like 'Take the second turning on the left'. 'Take the first exit from the roundabout', or 'Follow the signs to the motorway'.

A possible extension of the system to incorporate speech recognition is also being considered. The driver would then simply be able to get in the car in the morning and say 'Go to the office' and be given the best route avoiding traffic jams.

Renault, which is the only company currently selling cars with speech chips, has carried out some studies of driver reaction times with visual signals. These suggest that the percentage of drivers responding to a flashing signal on the dashboard reaches only 70% after three seconds, and for some drivers it takes minutes or hours – they simply do not see the signals with their attention on the road. If the signal is an important one, even a three-second delay could be serious. Speech signals, on the other hand, generally produce an immediate response.

What clearly must be avoided is the intrusion of speech signals when they are unwanted – it should always be possible for the user to control what is to be heard, if anything.

Other applications that are coming into use and take advantage of the power of speech to give warnings that are heeded include talking petrol pumps which warn the motorist who is about to put diesel fuel into the tank. More refined versions could include a variety of fuels, telling the driver whether, for example, leaded or unleaded petrol has been selected. An application which could enhance safety is a spoken warning when buses or trucks are reversing, instead of the usual bleeped warning. The message can be something like 'Warning: bus reversing' and needs to be in clearly understandable speech. All of these are applications in which the speech systems laboratory of Texas Instruments in the UK has been involved.

SPEECH CODING

New opportunities in speech input and output have resulted from advances in speech coding – the compression of speech messages so that they can be stored more easily and transmitted digitally over busy telephone lines. Low-cost telecommunications chips can be used for coding messages for subsequent playback in voice mail and similar applications. Texas Instruments recently demonstrated a very simple device with two storage chips which could be used, for example, alongside a house door-bell. It could play back a message to the effect that the householder was not available at that moment and asking the caller to leave a message; then it could store the spoken message for subsequent recovery by the householder.

Similar technology is being considered for use with the proposed second generation standard cordless telephones (CT2) which will operate within a few yards of a public 'call box' using a 900-MHz radio link – not needing a long external aerial. The very compact cordless handset could incorporate a voice coder consisting of a digital signal processor using adaptive delta pulse code modulation (ADPCM) for coding the voice, together with an industry standard codec chip.

APPENDIX: SYSTEMS AVAILABLE

THE FOLLOWING information is based on replies to a question-
naire sent to companies in the business of speech technology. It
makes no claim to completeness, not least because replies were not
received from some companies. It does, however, include most of
the leading speech technology companies in the USA, UK and
Europe.

AT&T Conversant Systems

Manufacturer, distributor and systems integrator.

Address: 6200 East Broad Street, Columbus, Ohio 43213, USA
Tel: (614) 860 2000
Contact: Harry M. McHugh, Marketing and Sales Director.
Business sectors served: Voice response market.
Products/services:

Conversant 1 Voice System is a telephone voice response system for computer database access using touch-tone or voice input. It can process up to 32 calls simultaneously on one system, and systems can be networked. It is available in two models – Model 32 and Model 80. The latter is designed with increased built-in redundancy and back-up features, offers greater disk capacities and can support communication to multiple host computers. Capabilities include:

- Voice response using sub-band coding at 16 or 24Kbits/s, or multi-pulse linear predictive coding at 9.6 and 14Kbit/s.
- Speech recognition with isolated or connected-work input. The system offers speaker-independent recognition of digits 0–9, 'yes' and 'no' embracing regional accents and dialects from throughout the USA.
- Voice coding for compression and storage using sub-band coding.
- Speaker verification. The system uses voice print analysis to check speaker's claimed identity.
- Call-classification analysis of outward calls. If line is busy it dials again later. If phone is disconnected it notifies the database.

Autophon

Distributor and systems integrator.

Addresses: Headquarters in Switzerland: Autophon Telecom Ltd, CH-4500 Solothurn, Ziegelmattstrasse 1–15
Branches in Austria, Belgium, West Germany, France, Italy, Norway, Netherlands, UK. UK Systems Division: Autophon (UK) Ltd, 225 Frimley Green Road, Frimley Green, Camberley, Surrey GU16 6LD, UK
Tel: 0252 836776 *Telex:* 858304 *Fax:* 0252 836776
Contact: Joe McHugh, Director
Business sectors served: At-home services – home banking, shopping etc.; financial services; retail and distribution.
Products/services:

Voice announcement, voice response and voice messaging systems. The group distributes the US Periphonics TalkBack system which allows any telephone to be used to enter and receive data from a central computer and to perform real-time transactions. Voice response is digitised but not compressed, giving high-quality speech output. Input is through touch-tone keypad, and if the user does not have multi-frequency telephones a separate tone-pad can be provided to generate the tones. The system will link either directly with a mainframe computer or will run off-line for subsequent batch processing. It is assembled into standard 19-in. racks with from 13 to 104 slots. Cards for these slots are the TalkBack processor, host interface, telephone line interface and vocabulary storage. Line interfaces and software are also available to support Viewdata terminals including the French Minitel. Each vocabulary board can hold up to 32 seconds of continuous speech, corresponding to 40–50 words. Up to 19 vocabulary storage boards can be accommodated in one system.

Berkeley Speech Technologies Inc.

Developer and supplier of software. Systems distributor and integrator.

Address: 2409 Telegraph Avenue, Berkeley, California 94704, USA
Tel: (415) 841 5083
Contact: Elisabeth W. Peters, General Manager.
Business sectors served: Audio telephone information entry and retrieval.
Products/services:

BeSTspeech Integrated Telecommunication System allows users to communicate with computerised information using any telephone as a remote audio terminal. Speech output can be any mix of digitally-recorded human voice at a variety of bit rates, and computer speech synthesised in real time from ASCII text by Berkeley Speech Technologies' T-T-S proprietary text to speech process. Input methods include touch-tone recognition from the keypad, digital voice recording of incoming calls, and voice recognition.

The company builds turnkey systems to customer specifications, providing modular software integration of various kinds of standard hardware. Developers' packages offering various board and software combinations using BST's own toolkit are also available for those who wish to do their own system integration. Information bases and application programs can reside on mainframes, minis or micros as appropriate. T-T-S is also licensed as software to manufacturers for use in other kinds of products. It was developed for the 8088 series microprocessors and the TM320 signal processing chips, but can be ported into many other kinds of hardware environments.

Bridge Speech Systems

Manufacturer of Automated Operator call processor, and voice response systems.

Address: 77 West Las Tunas, Suite 202, Arcadia, California 91006, USA
Tel: (818) 447 9425
Contact: Ron Emerling, President.
Associated with Voice Technology Inc., *q.v.*

Cambridge Computer Solutions

Manufacturer and distributor of computer products.

Address: First Floor, 64 Cherry Hinton Road, Cambridge CB1 4AA, UK
Tel: 0223 213374
Telex: 817056 *Fax:* 0223 410291
Contact: Paul Walters.

Speech product:

Type and Talk Speech Computer operates in text to speech mode or in 'phoneme mode', receiving ASCII text through a Centronics or RS 423 interface. It has a Z80 microprocessor with 8k byte of ROM and 2k byte of RAM, connected to a Votrax SC-01A speech chip. Speech output is amplified and spoken over an internal 8-ohm speaker. There are controls for volume and speed, and a socket for external earphone or speaker. In text to speech mode all the text sent to the computer will be spoken using built-in software which converts to text into the phonemes required by the speech chip. In phoneme mode the text received is expected to represent phonemes, and the speech chip may be programmed directly. The computer can say any number with up to 12 digits after the decimal point and many digits preceding the point. Pound, dollar signs and many mathematical and other characters are spoken correctly, and the computer deals intelligently with hyphens and full points.

Computer Gesellschaft Konstanz mbH

Manufacturer and systems integrator.

Addresses: Max-Strohmeyer Strasse 116, D-7750 Konstanz, Federal Republic of Germany

Tel: 07531 874218 *Telex:* 733332 *Fax:* 07531 874567
UK: Siemens Ltd, Siemens House, Sunbury-on-Thames, Middlesex
TW16 7HS, UK
Tel: 09327 85691 *Telex:* 8951091
Siemens offices in other countries.
Contact (at Konstanz); Peter Gütinger.
Business sectors served: End users and OEMs.
Products/services:

Speech recognition system Computer Speech Entry CSE 1200 is in use in industrial environments with a recognition accuracy claimed to be better than 98%. It consists of a perceptually orientated speech preprocessor which extracts the required data from microphone input and presents it as a digital stream to a 16/32-bit microprocessor running at 10MHz. The microprocessor is programmed as a single-utterance speaker-dependent recogniser. The nominal vocabulary limit is 500 words. Response times are under 300 milliseconds and vocabularies can be segmented. Recognition performance is measured as 99.48% with the industry standard Texas Instruments database. A prompt and output string of up to 16 ASCII characters can be defined for every word. Training repetitions can be set between 3 and 10 times, and the words are presented in random sequence.

The company is able to advise on suitable applications; install recognition systems; integrate them with existing data systems; create complete systems solutions; and carry out maintenance and training. Industrial applications to date include incoming and outgoing goods inspection; warehousing and inventory control; quality control/assurance; package and parcel sorting; CAD input; screen control for data display; and automation control in the factory.

Denniston & Denniston Inc.

Software supplier and integrator of complete speech systems.

Address: 3250 N. Arlington Heights Road, Arlington Heights, Illinois 60004, USA
Tel: (312) 398 8500
Contact: Wm B. Denniston Jr, President.
Market sectors served: Horizontal market with speech application toolkit. Vertical applications in health care, banking and manufacturing.

Products/services:

SofTalk II is a set of speech application modules which can be configured in a number of ways for many different applications, based on the IBM PC or compatible. It can operate as a multi-user system for up to seven users and it can be supplied with a telephone interface and local area network options. Speaker-dependent recognition can be built up with vocabularies in blocks of up to 50 utterances in the active vocabulary. A speaker-independent recognition facility is provided for the numeral digits, 'yes' and 'no'. The linear predictive coding method used in recognition and synthesis allows connected speech to be used, and vocabulary words can be recognised within a natural sentence. The single-user version can convert text to speech within a dialogue, making it suitable for voice messaging and voice response database applications as well as such things as industrial quality control.

The company also offers a complete range of support services including installation, training and startup, together with update facilities.

Digital Equipment Corporation

Manufacturer and supplier of DECtalk equipment and software.

Address: 146 Main Street, Maynard, Massachusetts 01754, USA. ML03–3/48
Tel: (617) 493 3587
Contact: Edward Lazar, Manager, Voice Products Group or any Digital sales office throughout the world.
Products/services:

DECtalk text to speech system converts standard ASCII text into human-quality speech. It offers a choice of male, female and child voices, has a variable speaking rate from 120 to 150 words per minute. It has two pronouncing dictionaries, for standard and user-specific words, together with a library of letter-to-sound rules enabling it to generate exact pronunciations for more than 20 000 words. It accepts input from any computer via a standard RS 232C port, and from a touch-tone keypad. Voice output goes to built-in speaker, headphones, audio jack or telephone.

Dual-line and multi-line DECtalk are designed for medium and high volume multi-user communication systems, and the DECtalk board is a single module enabling system designers to integrate DECtalk workstations or telephone response systems.

DECtalk Voice Response System can be used as stand-alone or as front end to a database, and to manipulate data independent of the database. It employs a dual-line DECtalk unit, a MicroVAX II microcomputer with fixed disk drive, tape drive, communications controllers and video terminal. The entry level system has two lines, and upgrades can expand this to 32 lines. An Application Development Guide helps users to develop their own voice response applications.

Digital Sound Corporation

Manufacturer of voice messaging, response and recognition equipment.

Address: 2030 Alameda Padre Serra, Santa Barbara, California 93103, USA
Tel: (805) 569 0700   *Telex:* 530133   *Fax:* (805) 586 0098
Contact: David E. Meldrum-Taylor, Director of International Marketing. International distributors include Olivetti of Ivrea, Italy, and Mitel of Kanata, Ontario, Canada.
Products:

VoiceServer 1000, 1500 and 2000 are telephone voice applications processors incorporating voice mail, voice response and call processing. An automated attendant answers calls promptly and permits the caller to dial an extension, wait for the operator or leave a message. If an extension is busy or nobody answers, the system informs the caller and provides options. The larger 2000 sytem can hold 3000 mailboxes and handles up to 20 simultaneous users. Other options include facilities to assist transcribing voice messages using voice commands, voice-prompted questionnaires and application forms, and a guest messaging system for hotels. The Voice Development Server is a software toolkit designed to aid in the development of customised voice applications. A text to speech option is available, which can allow such facilities as telephone playback of electronic mail messages or newswire information. There is also a speech recognition option offering speaker-independent recognition of up to 32 isolated utterances per vocabulary.

First Byte

Manufacturer of text to speech synthesis equipment.

Address: 3333 E. Spring Street, Suite 302, Long Beach, California 90806, USA
Tel: (213) 595 7006
Contact: Richard Jacks, President.
Business sectors served: Education, handicapped, industrial, office automation, military.
Products:

SmoothTalker is a text to speech synthesiser developed using the company's own algorithms and claimed to be the first to operate entirely in software on all major microcomputers, including Apple, Atari, Commodore, IBM and Tandy. It has two separate sections, each of which need only be resident in memory while words are being spoken. Most applications will swap sections as needed. A 'front end' takes normal English text and converts it into 44 phonemes. It also applies over 1200 English rules to the incoming text and automatically encodes stress, pitch, inflection and so on caused by punctuation. The 'back end' converts phonemes into speech through the speaker. For issuing a prestored message to the user, only this 'back end' must be memory-resident, and only when the actual speech is taking place.

The company also has a range of speech synthesis products for the educational market, covering spelling, writing and elementary mathematics for children between 3 and 12 years old. These products are distributed by Electronic Arts, PO Box 7530, San Mateo, California 94403, USA.

IOCS Inc.

Systems integrator, also manufacturer of Voice-Net voice response system.

Address: 400 Totten Pond Road, Waltham, Massachusetts 02254, USA
Tel: (202) 879 7000
Contact: Maureen Halley Lederhos, Marketing and communications specialist.
Business sectors served: Banking and financial services, transportation, health care and insurance.
Products/services:

Voice-Net is an interactive voice response system using high quality digitised human speech in a vocabulary expandable to 5400 words and phrases. It uses touch-tone telephone input and will support from 2 to 64 callers simultaneously. It can interact with the user's existing computer or operate in a stand-alone mode. Applications generation software allows new or revised applications to be generated without the need for programming. Periodic statistical reports and an audit trail of transactions are available. The system includes a proprietary database for information storage. Systems installed by IOCS include a voice information network for aviation weather reports, and an interactive voice response system providing a variety of services to customers of the Bank of Boston.

Interstate Voice Products

Manufacturer, distributor and systems integrator.

Address: 1849 West Sequoia Avenue, Orange, California 92668, USA
Tel: (714) 937 9010 *Telex:* 47–22046 *Fax:* (714) 758 3222
Contact: Peter C. van der Most, Director of Marketing
UK distributor: Kode Computers Ltd, Drakes Way, Swindon SN3 3JL, UK. *Tel:* 0793 511345 *Telex:* 449335 *Fax:* 0793 511467; distributors in Australia, Hong Kong, Italy, Switzerland, Taiwan, UK and West Germany.
Business sectors served: Cellular telephone industry, factory/office automation markets.
Products/services:

Vocalink range of voice recognition products. VRC100-2A is a two-chip set for use as building blocks in speech recognition systems, with 100 or 200-words vocabulary. Input is analysed by a 16-channel spectrum analyser. EPROM firmware accommodates 17 user commands including two training commands. Recognition accuracy is better than 98.99% on test vocabularies in 100-word subsets. VRT300 is a single board recognition module for DEC VT100 and other terminals. CSRB is a connected speech board for IBM PC/XT/AT, with a 240-word vocabulary, with optional Texas Instruments voice synthesis module. SRB-LC is a voice recognition package for IBM and compatibles with isolated word vocabulary of 500, grouped in menus of up to 20 words. SYS300 is a self-contained voice recognition system with 200-word vocabulary, which does not require special programming to interface to most ASCII terminals. It has an on-board editor for vocabulary training. Series 4000 includes a voice of 100 words. It includes PC-DOS or MS-DOS software tools to define recognition grammars and recognition output messages. Optional is a text to speech response facility. There is also a Series

4000 Voice Planner with a continuous speech applications development software package. The company also produces a Cellular Module to provide voice input dialling other facilities for in-car cellular telephones.

Kurzweil Applied Intelligence Inc.

Manufacturer of voice to computer reporting and dictation system.

Address: 411 Waverley Oaks Road, Waltham, Massachusetts 02154, USA
Tel: (617) 893 5151 *Telex:* 9103809408
Contact: Robert Joseph, Vice President of Marketing.
Business sectors served: Currently concentrating on medical applications.
Products:

VoiceRAD is a 1000-word speech to text system designed for doctors to dictate, edit and print radiology reports in seconds. Besides allowing a large isolated-word specialist vocabulary, it can 'trigger' words or phrases to send complete regularly-used sentences or paragraphs into the report. Each word appears on a video screen as it is spoken into the microphone, and once the report is ready, the radiologist tells the system 'Print this'. A larger version, VoiceRAD-MX (Radiology Multiple Examination), has a vocabulary which can be expanded to 10 000 words. VoiceEM is a similar system designed for emergency departments in hospitals to generate reports in less than a minute using predefined complaint and physical examination tests, ancillary test results, and diagnoses and discharge plans. The core vocabulary contains the words and phrases to generate reports covering more than 20 complaints, and it can easily be modified to include others.

Logica

Systems integrator, distributor and manufacturer.

Address: 64 Newman Street, London W1A 4SE, UK
Tel: 01-637 9111 *Telex:* 27200 *Fax:* 01-637 8229
Contact: Dr Jeremy Peckham, Manager, Speech and Language Division.
Business sectors served: All, but with emphasis on industry, defence, commerce, finance and telecommunciations.

Products/services:

Industrial implementations by Logica have been performed for clients including Jaguar Cars and Caterpillar. The company places particular emphasis on using speech-based data capture as a highly versatile tool within integrated manufacturing systems. In 1981 Logica developed the world's first continuous speech recognition system, Logos, capable of processing very large stored vocabularies of thousands of words arranged in up to 700 active vocabularies. The technology has been licensed to Racal Acoustics and Smiths Industries and a small number have been sold to government and commercial research laboratories. Recently developed is a smaller version, Logos 286/20, with a capacity of 600 utterances in up to 30 active vocabularies.

Loughborough Sound Images Ltd

Manufacturers of speech and digital signal processing products. Also undertakes design contracts and consultancy work to customers' specific requirements.

Address: The Technical Centre, Epinal Way, Loughborough, Leicestershire LE11 0QE, UK
Tel: 0509 231843 *Telex:* 341409 *Fax:* 0509 262433
Contact: Dr D. J. Quarmby, Technical Director, or S. Yates, Design Engineer
North American distributor: Spectrum Signal Processing Inc., 460 Totten Pond Road, Waltham, Massachusetts 02154, USA
Products:

Speech recognition and synthesis board plugs into IBM PC/XT/AT or compatible for design of speech interaction systems. It uses the PC for storage and control and the specialised processors for speech pattern analysis, matching and synthesis: speech analyser provides 16 band-pass filters with energy measurement per frame; pattern comparator contains two high-speed computers, one performing distance measurements between frames, the other performing dynamic time warping algorithm; speech synthesiser independent subsystem is controlled at a detailed phonetic level, using the company's implementation of the parallel formant model. A speech synthesis module occupying less space than a credit card contains a microcomputer to provide interface between the user environment and the synthesiser, with firmware which can be tailored to user requirements. It offers three levels of synthesis – stored speech, phonetic synthesis and text to speech. A phonetic synthesis is claimed to give high-quality British English output with natural intonation controllable by the user. It comes with its own voice but can be tailored to other accents. The synthesiser chip embodies a parallel formant speech model developed at the Joint Speech Research Unit, and the phonetic control program is a rule-based scheme also derived from JSRU research. The company also markets a range of other digital signal processing products.

Marconi Speech and Information Systems

Manufacturer, distributor and systems integrator.

Address: Browns Lane, The Airport, Portsmouth, Hampshire PO3 5PH, UK
Tel: 0705 674180 *Telex:* 869442
Contact: David A. Guyatt, Sales/Marketing Manager
Business sectors served: Diverse areas from financial information services through industrial speech recognition systems.
Products/services:

Marconi has been in speech processing since 1966 and speech recognition since 1979. Macrospeak is a speech recogniser which is inserted in the serial link between host computer and keyboard, allowing mixed spoken and keyed commands with no change to the user's software. It incorporates a macro facility enabling the user to employ simple speech commands to initiate host computer tasks which previously required large amounts of command text. The system is trained by typing each word and its syntax menu name, and speaking it once. Syntax can be designed to prohibit invalid word sequences. Output from each recognised word sequence is a macro which controls an activity of the host computer. Vocabulary size is 160 or 640 words. Maximum duration for stored speech is 205 seconds, maximum number of syntax nodes 200, maximum number of macros 800. For defence applications is the ASR1000 Airborne Speech Recogniser, using concepts from Macrospeak with additional algorithmic improvements to give the high performance required by the Ministry of Defence.

Marconi also builds a range of announcement and voice response equipment. Systems include Marcall, a 30-channel telephone announcement system offering up to 7.5 hours announcement storage; Incall, a 60-channel direct dial-in equipment providing a 'start from the beginning' message from a 48-minute storage facility;

Automatic Change Number Interception equipment, allowing up to 255 callers to hear high-quality 'start at the beginning' change number announcements. An interactive voice response system using Marconi's Keycall system is the Financial Times Cityline service for investors.

Natural Microsystems Corp.

Manufacturer of Watson voice mail and related equipment and software.

Address: 8 Erie Drive, Natick, Massachusetts 01760-1313, USA
Tel: (617) 655 0700 *Telex:* 6502240 143MCI *Fax:* (617) 655 0700 X52
Contact: Bob Panoff, Vice President Marketing.
Business sectors served: All, particularly financial services and distribution.
Products:

Watson is a voice/data modem board which goes into an expansion slot of an IBM PC/XT/AT or compatible, allowing incoming telephone calls to be answered, messages to be sent in response to touch-tone signals, and incoming messages to be stored on disk in separate 'mailboxes' for each caller. It includes a 500-name telephone directory, autodialler, 300/1200 baud modem, call forwarding, time tracking, an electronic calendar that automatically calls users to remind them of appointments, dictating capability, and the ability to use information in its database from any touch-tone telephone. There are application packages designed for individual types of businesses. Watson VBX is an IBM PC-based voice-messaging and telephone-management system for businesses with up to 100 employees. Watson Quickspeak is a text to speech system for providing information to authorised people with a touch-tone telephone. It identifies and loads the appropriate text to be spoken. Speech is created from stored text files and keyboard input. Speech is interpreted using linguistic and phonetic rules. Different text types such as memos and catalogues have different linguistic conventions, and there is a moderately sized exception dictionary. The system provides context-sensitive word recognition and pronunciation, abbreviation expansion, interpretation of punctuation marks and pronunciation of cardinal and ordinal numbers.

PA Technology

Systems integration, contract design for manufacturers. The organisation is part of PA Consultancy Services Ltd.

Address: Cambridge Laboratory, Melbourn, Royston, Hertfordshire
SG8 6DP, UK
Tel: 0763 61222 *Telex:* 81451 *Fax:* 0763 60023
Contact: Nigel Sedgwick, Marketing Manager, Defence.
Business sectors served: All sectors, particularly product and process
development for manufacturing industry.
Products:

PA Technology has developed under sponsorship of the British
Technology Group – BTG – the ASR personal speech recogniser as a
low-cost high-performance chip set. BTG is currently (1988) seeking
licensees and partners to commercialise versions of the basic proven
technology for different markets. The module provides speaker-
dependent recognition of a 64-word active vocabulary, with rapidly
transferable vocabularies, giving continuous recognition using
algorithms based on Hidden Markov Modelling. BTG is at 101
Newington Causeway, London SE1 6BU, UK; tel: 01 403 6666.

PA technogy has also been working in-house on text to speech
synthesis and has developed vocoders for military and civil use.

Philips Speech Synthesis Products

Manufacturer of speech synthesis products.

Address: Mullard Ltd, Millbrook Industrial Estate, Southampton,
Hampshire SO9 7BH, UK
Tel: 0703 702701 *Telex:* 47640 *Fax:* 0703 701004
Contact: Michael Roberts, International Product Marketing Manager.
National distributors: UK: Mullard Ltd, Mullard House, Torrington
Place, London WC1E 7HD. *Tel:* 01 580 67633. *Telex:* 264341. *Contact:*
Mr B. Cowle; *USA:* Signetics Corp., 811 East Arques Avenue, PO
Box 3404, Sunnyvale, California 94008-3409. *Tel:* (408) 991 2000.
Contact: Matt Robison, Marketing.
Business sectors served: Telephony, automotive, PC, Education.
Products:

Philips produces a complete range of speech products to support
the PCF8200 formant speech synthesiser. This uses five formant
frequencies for male speech, four for female speech over 5-kHz
bandwidth. When not speaking it can be made to power-down
automatically, drastically reducing current consumption. The
OM8210 speech adaptor box analyses incoming speech and generates codes suitable for the synthesiser. It includes sophisticated
on-screen editing facilities and works with IBM PC/XT/AT and

compatibles. OM8200 is a low-cost demonstration/OEM Eurocard with speech synthesiser, audio amplifier and EPROM capacity for 12 minutes speech. External interfacing is provided via an 8-bit bus.

Racal Acoustics Ltd

Manufacturer, distributor and systems integrator.

Address: Beresford Avenue, Wembley, Middlesex HA0 1RU
Tel: 01 093 1444 *Telex:* 926288 *Fax:* 01 903 1253
Contact: Dr Peter Wheeler, Marketing Director.
Business sectors served: Industrial and defence communications, telecommunications.
Products:

A wide variety of electro-acoustic communciations equipment for aviation, defence and commercial applications. The range now includes avionics speech recognition systems and stored voice devices.

Smiths Industries Aerospace and Defence Systems

Manufacturer and systems integrator.

Address: Bishops Cleeve, Cheltenham, Gloucestershire GL52 4SF, UK
Tel: 024267 3333 *Telex:* 43172
Contact: Michael R. Taylor, Head of Speech Technology.
Business sectors served: Aerospace and defence.
Products/services:

Work has focused on development of a rig for research into the use of speech recognition for control of a control display unit connected to an aircraft navigation and management system, using continuous connected speech. The work is outlined in Chapter 4.

Speech Design GmbH

Manufacturer of speech synthesis and recognition equipment.

Address: Landsberger Strasse 33, 8034 Germering, Federal Republic of Germany
Tel: 089 84931-0
Contact: H. J. Scheele, Sales
Distributors: UK: STC-Mercator, South Denes, Great Yarmouth, Norfolk NR30 3PX, UK. *Tel:* 0493 844911; *Netherlands:* Maldius,

Schiedam; *Finland:* OY Ferrado, Helsinki; *Italy:* ATS, Rome; *Switzerland:* Fabrimex, Zurich.
Business sectors served: Industry, telecommunciations, computers, telemetry.
Products:

Speech synthesis development system SDM 300-T produces compressed speech output to an EPROM using the CVSD algorithm at a bit rate selectable between 17 and 43kbits/s. It supports generation of structured vocabularies with phrases stored once and used in multiple messages. ISAS and ISAS-P are intelligent speech output stations using EPROM vocabulary memories. ISAS is also available with speech input. Speech System II is an industrial speech recognition system which responds to verbal commands by operating 24 relays. Spoken prompts are provided by customised EPROMs generated with the SDM 300. Speech System III is an industrial speech input/output system, interfacing to most host computers through an RS232-C port. Basic recognition vocabulary is 128 isolated words, expandable to 80 times as many, and divisible into subsets. Prompts and feedback are provided by customised EPROM. Receptel 2000 is a hotel wake-up call system using its own keyboard input and synthesised speech output in three or more languages.

Speech Plus Inc.

Manufacturer of voice mail and database equipment.

Address: 640 Clyde Court, PO Box 7461, Mountain View, California 94306, USA
Tel: (408) 964 7023 *Telex:* 256684 *Fax:* (415) 961 3420
Contact: Carl L. Berney, Vice President, Systems Engineering
Sales are direct to end users through value-added resellers. Distribution in Canada is through Speech Dynamics Ltd, Toronto, Ontario.
Business sectors served: Telephone companies, insurance companies, railways, financial organisations.
Products:

The CallText range provides voice gateways that let end users access any database over the telephone. Primary applications include operator replacement or bypassing in order entry or enquiry, verification, database query, name and address locator services and field service dispatch. Messaging applications include electronic mail access by telephone and reading text-based messages from newswires. CallText 5100 is a programmable system supporting up to 7 channels simultaneously. CallText 500 is a module for IBM PC and compatibles. CallText 5050 is a multi-channel RS232-C peripheral operating under control of any host computer. The text to speech facility uses the company's own synthesis by rule algorithms. It accepts ASCII text and phoneme coded input and includes three male voices, word or clause prosody, intonation and phrasing controls, alternative word pronunciations and other functions.

Speech Systems Inc.

Manufacturer of phonetic-based speech recognition system dealing direct with end users.

Address: 18356 Oxnard Street, Tarzana, California 91356, USA
Tel: (818) 881 0885 *Telex:* 287282
Contact: Deana Murchison, Manager, Marketing Communications.
Business sectors served: Artificial intelligence, aerospace, finance, military, medical.
Products:

SSI has developed its own approach to speech recognition based on the analysis of 40 phonemes in speech sound. The Phonetic Engine accepts 'natural speech' and converts it into phonetics. The Phonetic Decoder software runs in a general-purpose computer and translates the phonetic codes into word strings. Speech Input Development Systems are complete operational speech recognition systems with the software tools for integrating input into existing or new applications. Besides the phonetic engine and decoder it includes a master dictionary of 12 000 English words, with phonetic and alphabetical spellings, and development toolkit software.

Texas Instruments

Address: PO Box 2909, Austin, Texas 78769, USA
Tel: (512) 250 7111
Address UK: Regional Technology Centre, Manton Lane, Bedford MK41 7PA, UK
Tel: 0234 270111 *Telex:* 82178
Products/services:

Products include the TI-Speech system, incorporating recognition, synthesis and analysis, with text to speech and telephone management capabilities. It is board-mounted for use with either a member of the TI Professional Computer family or an IBM PC or compatible product. The TMS50C40 is a programmable synthesiser/ROM integrated circuit for speech generation, using a Linear Predictive Coding algorithm. The company provides a speech synthesis service for large-scale users.

Voice Industries Corp.

Manufacturer of voice recognition and response equipment.

Address: 185 Ridgedale Avenue, Cedar Knolls, New Jersey 07927, USA
Tel: (201) 267 7507 *Fax:* (201) 267 0548
Contact: John Ferretti, Vice President Sales/Marketing.
International distributors: UK: Decade Computers Ltd, Ringway House, Kelvin Road, Newbury, Berks RG13 2BD, UK. *Tel:* 0635 38008 *Telex:* 846301 *Fax:* 0635 521268; *Contact:* Steve Bromley, Technical Support Manager.
Business sectors served: Industrial, military, aerospace, financial, medical.
Products:

Verbex Series 5000 voice I/O system adds voice recognition and response capabilities to personal, micro, mini and mainframe computers via RS23-C asynchronous ASCII input/output ports at up to 19 200 baud. It provides continuous, speaker-dependent recognition with a 100-word vocabulary, and voice response through a text to speech facility with output generated within the Series 5000 or by the host computer. All the voice application information and user patterns are contained in a CMOS cartridge which is easily changed for different user applications. The system uses multiple grammars to define vocabularies and word sequences and to limit the number of valid phrases expected at any time. A 'continuous listening' capability eliminates the need for manual control of the microphone by use of proprietary software. Voice Developer is a menu-driven MS-DOS software program for developing customised voice applications by using a PC with an asynchronous port. It allows creation of grammar, verbal responses and voice prompting, and a translation table to generate control codes in the computer.

Voice Systems International Ltd

Systems integrator.

Address: Cambridge Science Park, Milton Road, Cambridge CB4 4GF
Tel: 0223 862327 *Telex:* 8950511 *Fax:* 0223 861898
Contact: Chris Booker, Managing Director.
Business sectors served: All sectors.
Products/services:

VSI carries out integration of complete systems involving voice recognition and response, with particular experience of industrial inspection and other applications and telecommunciations systems involving multiple-line telephone access with voice recognition and response. The company uses equipment most suitable for each application and has extensively used Votan equipment (see below). Implementations at Austin Rover, Rolls-Royce and Virgin Group are described elsewhere in this book.

Voice Technology Inc.

Distributor and systems integrator of call processors, voice mail and voice response systems.

Address: 3010 LBJ Freeway, Suite 1515, Dallas, Texas, USA
Tel: (214) 243 6366
Contact: L. Northam, Vice President
Products:

VTI Call Processor comes with up to five minutes of message storing using digital recording to dynamic RAM. It provides direct dialling of extensions and two levels of menus accessed by DTMF tele-phone. Callers have options of alternative extensions, and voice mailbox facilities, and calls can be automatically forwarded to individual voice mailboxes. Company is associated with Bridge Speech Systems, *q.v.*

Votan

Manufacturer and systems integrator.

Address: 4487 Technology Drive, Fremont, California 94538, USA
Tel: (415) 490 7600 *Telex:* 176274
UK distributor: Voice Systems International Ltd (see above).
Products/services:

Votan was founded in 1979 to develop and market voice technolo-gies and has installed thousands of systems in major companies in the USA. Products include the Voice Card, a plug-in for the IBM PC and compatibles, with continuous voice recognition and voice recording and playback. Software packages for it include the Telephone Professional 'secretarial' functions, VoiceKey voice-activated keyboard macro functions, and software development tools. A voice terminal provides similar capabilities to a variety of microcomputers and minicomputers as well as to IBM 3270 networks. TeleCenter is a stand-alone voice mail system for organisations with 25 to 250 employees. It is based on the IBM PC XT and will record, play back and distribute individualised messages for all callers, using speaker-independent voice recognition with a 21-word vocabulary, or touch-tone keypad. There is also a stand-alone multi-channel voice processing and telephone management system for applications with a large number of users.